Speed and the Thoroughbred
The Complete History

Speed and the Thoroughbred
The Complete History

by Alexander Mackay-Smith

Illustrated with
Paintings, Prints and Photographs
of the Breeders, Stallions and Mares
who Shaped the Thoroughbred,
from First Century Irish Hobby
to Twentieth Century Secretariat 1970

With a Foreword by John T. von Stade

2000
The Derrydale Press
Millwood House

THE DERRYDALE PRESS

Published in the United States of America
by The Derrydale Press and Millwood House, Ltd.
4720 Boston Way, Lanham, Maryland 20706

Distributed by NATIONAL BOOK NETWORK, INC.

Copyright © 2000 by Alexander Mackay-smith
First Derrydale printing 2000

All rights reserved. No part of this publication may be reproduced,
stored in a retrieval system, or transmitted in any form or by any
means, electronic, mechanical, photocopying, recording, or otherwise,
without the prior permission of the publisher.

British Library Cataloguing in Publication Information Available

Library of Congress Cataloging-in-Publication Data

Mackay-Smith, Alexander.
    Speed and the thoroughbred : The Complete History / Alexander Mackay-Smith.
       p. cm.
    Includes bibliographical references.
    ISBN 1-58667-040-9 (cloth : alk. paper)
    ISBN 1-56416-192-7 (leather : alk. paper)
    1. Thoroughbred horse—Great Britain—History. 2. Race horses—Great Britain—Speed—History. I. Title.

SF293.T5 M26 2000
636.1'32'0941—dc21
                                                  00-043109

The paper used in this publication meets the minimum requirements of
American National Standard for Information Sciences—Permanence of
Paper for Printed Library Materials, ANSI/NISO Z39.48-1992.
Manufactured in the United States of America.

PATRONS

Mr. and Mrs. Charles C. Abeles
Damaris S. Abeles and Brown Sharp II
Jessica A.K. Abeles-Wong and José N. Wong
Mr. and Mrs. Nathaniel C. Abeles
Arthur W. (Nick) Arundel
Mr. Robert L. Banner, Jr.
Mr. and Mrs. W. Bell-Watkins, IV
Dr. Deb Bennett
Mr. and Mrs. Edward S. Bonnie
Edward L. Bowen
Elizabeth Anne Carnes
Phil and Margot Case
Mr. and Mrs. Rustom A. Cowasjee
W. Joan Mackay-Smith Dalton and Family
James M. and Emily Mackay-Smith Day
Mrs. Lynne Kindersley Dole
Norman and Joan Fine
Mr. and Mrs. Bertram R. Firestone
Mrs. Helen K. Groves
Andy and Mary Scott Guest
Jacqueline and Edward Harris
Dr. and Mrs. E.C. Hart
Damaris S. and John R. Horan
Mr. and Mrs. George A. Horkan, Jr.
Indian Hills Equestrian Center
Matthew Klein
J. Edward and Caroline Treviranus Leake
Toni Lee
Mason and Mary Lu Lampton
Mrs. Stacy Lloyd
Mr. and Mrs. Alexander Mackay-Smith, Jr.

Amanda Joan Mackay Smith
Catharine Mackay-Smith and Kenneth Kempson
Justin and Meridith Mackay-Smith
Matthew and Wingate Mackay-Smith
Marion duPont Scott Equine Medical Center
Winslow and Juliet Mackay-Smith McCagg
Dr. and Dr. William H. McCormick
Mrs. G.W. Merck
Jerry L. Miller
Mr. and Mrs. George L. Ohrstrom, Jr.
James Oury
Robert F. Oury
Herbert J. Richman
Charlotte L. Robson, DVM
Dr. and Mrs. Joseph M. Rogers
Thomas Ryder
Lindsay and Edgar Scott
Mr. and Mrs. Ami Shinitzky
Mr. and Mrs. J. Donald Shockey, Jr.
Gordon G. Smith, ex-MFH
Patricia R. St. Clair
William Steinkraus
John A. Terry
Denya Massey Treviranus
Peter S. Tsimortos
Hedda Windisch von Goeben
Mr. and Mrs. John T. von Stade
Stella Walker
Hon. Charles S. Whitehouse
Mrs. Julian Whittlesey
Mildred Wilson

Dedicated to

Abram Stevens Hewitt (1902–1987)

Who shared with the author
a lifelong friendship and
fascination with the
Thoroughbred Racehorse

Acknowledgments

The acknowledged leader is Denya Treviranus. She has typed the entire manuscript from its conception, through its many phases, including the discovery of new information and revisions, with editorial skills, good humor and understanding over 10 years. Her comments and suggestions have been invaluable. The book could not have been written without her.

Many years ago, George Ohrstrom, Sr. and Jr., proprietors of the *Chronicle of the Horse*, urged the selection of this subject, and over the years continued to encourage me, recognizing the book's importance. The most noted supporter of scholarship in the field of sporting art, Paul Mellon, enabled me to write this book through his generous fellowship. His generosity included the loan of several important illustrations in this book, but it extended beyond, to a demonstrated desire to ensure this work was completed to take its place in the annals of the history of the Thoroughbred.

My fellow Harvard Law School graduate, Charles Calvert Abeles, contributed invaluable legal services, leadership and wise counsel in support of the author and this scholarly work. Hetty Mackay-Smith Abeles graciously provided much research and cheerfully unearthed critical pieces of information needed to write such an intensively researched book. Matthew Mackay-Smith provided information, listened enthusiastically, and considered all my new theories and conjectures—some included in this work.

Through their work on the *Dictionary of American Animal & Sporting Painters*, Turner Reuter, a fellow MFH and sporting art collector, and Joshua Mackay-Smith, have been able to add detailed information that would have otherwise been missing.

My dear friend, Stella Walker, the great English authority on sporting artists, provided a wealth of information, not simply for this book but for the author's many literary efforts. She has been generous and interested in all my projects through our many years of friendship.

Without Norman Fine, the book's publisher, the illustrations would have been insignificant. He took over the heavy burden of locating and collecting illustrations, helped with their titles and captions, gave generously of his time and friendship, and quietly but with determination moved the book toward its publication. Edward Harris, a long-time friend and collector of sporting art, underwrote a substantial portion of the expense of the illustrations, enabling the book to include some important pieces which might otherwise have been excluded.

The collections of the National Sporting Library, Middleburg, Virginia, and the members of its staff have provided vital assistance, particularly Peter Winants, Laura Rose and Judith Ozmont. Dr. Joseph P. and Donna Rogers understood this undertaking to be a scholarly work, representing years of research and compilation. Their determination to see it published was demonstrated by their unfailing support.

At the outset, Edward Bowen demonstrated his support, interest and faith in the author's early attempts to present new ideas and information about the Thoroughbred to the subscribers of the *Blood Horse*. As early as 1987, he published early drafts of some of the eventual text of this book. Acknowledgments must also be made to the libraries where the author searched for obscure pieces of information in unraveling the puzzles. Of special note is the Keeneland Library in Lexington, Kentucky and the British Library in London, England. The staff at both libraries were continually helpful and resourceful.

Several people read parts of the manuscript in detail during various stages of its development, providing valuable suggestions as to content and style. Particularly acknowledged is Caroline Treviranus Leake, who with great patience read much of the book to the author as it was typed.

Appreciation is also extended to John Terry of Marske, the seat of one of the important early studs in Yorkshire, England, for the photographs and historical information that he furnished. Historian Thomas Ryder of York, England, researched important details about this story and provided invaluable information to the author.

Acknowledgments

Finally, I am grateful to my wife for her understanding and good-humored tolerance of the many hours, indeed years, required to write this and the many other books over the past 33 years. Marilyn listened to every story, helped me unravel complex and often contradictory information, and journeyed with me through the maze of this complicated story.

Alexander Mackay-Smith
1998

Foreword

Since the turn of the century, members of my family have been deeply involved with the Thoroughbred. My father, the late F. Skiddy von Stade, was an amateur steeplechase rider who served as president of the National Steeplechase and Hunt Association in the early 1940s. He was also the president of the Saratoga Association (commonly known as the Saratoga Racing Association) from 1943 until 1955 when it merged into the Greater New York Association, forerunner of the New York Racing Association.

In my generation, several of my brothers and sisters, as well as in-laws, have enjoyed various connections with the sport, ranging from the pleasures and rigors of riding, training and owning steeplechasers and flat horses to involving ourselves with numerous organizations which support the turf. I dwell on this to illustrate that the Thoroughbred has been such an important part of our family, as it has to so very many others, through more years than we might like to contemplate.

The grace and beauty of this animal, and his competitive spirit and nobility, have long been recognized by all of us. This combination is just as potent today as it has been for centuries. Whatever one's length of association with the Thoroughbred, it is a particularly engrossing experience to learn more about the origins of this remarkable animal through the diligent research of Alexander Mackay-Smith.

For many, it has been an easy assumption that the Eastern bloodstock introduced into England was the taproot of the outstanding traits of today's Thoroughbred. After all, a certain romantic image is easily conjured of the Arab rider and his desert companion, the man dependent on the loyalty, tractability, endurance and speed of his mount. Well, Mr. Mackay-Smith adroitly sets us straight. True, Eastern imports had their place in the foundation of what we know as the Thoroughbred. The signature quality of speed, however, has quirkier origins.

One of the delights of the saga, as Mr. Mackay-Smith sets it out, is how early the relationship which we feel between ourselves and the Thoroughbred began to enter the hearts and souls of men. We might be tempted to think of Irish Hobbies, the horses which the author identifies with the "earliest sprinting speed" as somewhat archaic compared to the modern Thoroughbred. However, Mr. Mackay-Smith quotes the poet Paulus Jovius extolling their virtues as early as 1548:

> There is something magnificent, a kind of majesty in his whole frame, which exalts his rider with pride as he outstrips the wind in his course.

On a different level, in responding to the Irish Hobbies, King Henry VIII, a fellow known to have had a pragmatic approach, lamented the news that Ireland "produces…most excellent, victorious horses, more swift than the English horses."

Speed and the Thoroughbred Racehorse is the account of how "the three sources of Thoroughbred speed were finally combined—the two sprinting speed strains of the Hobbies and the English Running-Horses, and the middle distance speed strain of one unique horse, Place's White Turk. Of the three strains, only the middle distance speed strain was influenced by a new source: the importation (1730) and remarkable prepotence of the Godolphin Arabian more than half a century later.

It is more than appropriate that the individual whose research, determination and erudition finally circumscribed this history is Alexander Mackay-Smith. His life has been one of equal parts sportsman, historian, author and vigorous horseman. He has long been enthralled by the relationships of man, horse and hound, whether it be the stud of Cleveland Bays and Anglo-Clevelands he once developed, the companionable mounts of the Virginia hunt field, or the tumult of the race itself.

I have known Alex for some 35 years and have always admired his tenacity in researching his literary endeavors. His wealth of knowledge on any subject dear to his heart is unique. As a result, what we have in this volume is a productive confluence: a breed with a noble history entrusted to a well-qualified sportsman and scholar, who addresses his task with reverence.

John T. von Stade
July, 1998

Contents

    Patrons v

    Dedication vii

    Acknowledgments ix

    Foreword xiii

    Timelines xxii

    Taproot Pedigree of the Thoroughbred Racehorse xxvi

I   An Overview 1

II   The Irish Hobby 19

III   Hobby Influences in England 33

IV   Running-Horses 53

V   Sprinting Speed Strains Before 1650 71

VI   Perils and 'Saviors':
       Foundation Mares and Their Breeders After 1650 83

VII   Imported 'Pure' Arabians: *General Stud Book* Entries 95

VIII   A Century of King's Plates, 1665–1780 107

IX   Diplomatic Gift Stallions: Barbs and Turcomans 117

X   The Godolphin Arabian 1724–1753 131

XI   The Duke of Cumberland and Eclipse 151

XII   Surviving Speed Lines into the Twentieth Century 163

    Bibliography 175

    Glossary 179

    Index 185

Illustrations

Timeline of Events, People, Studs, and Horses xxii

Taproot Pedigree of the Thoroughbred Racehorse xxvi

Flying Childers by John Wootton 11

Eclipse with Mr. Wildman and His Sons by George Stubbs 15

Art Macmurchada, King of Leinster,
on a White Irish Hobby, 1399 21

Palio Race in Florence, 1418 24

Alfonso D'Este, Duke of Ferrara by Titian .25

King Henry VIII 27

Gerald FitzGerald, Ninth Earl of Kildare
and Viceroy of Ireland (school of Holbein) 29

Castle Mattress (now Castle Matrix),
Rathkeale, County Limerick, Ireland 30

Sir William Berkeley, Royal Governor of Virginia 31

Green Spring, near Jamestown, Virginia
(watercolor by Benjamin Latrobe) 32

Francis Manners, Sixth Earl of Rutland 35

George Villiers, First Duke of Buckingham
(painting attributed to Gerrit van Honthorst) 36

The Duke of Buckingham by Peter Paul Rubens 38

The Buckingham Family by Gerrit van Honthorst 39

Thomas, Lord Fairfax, Third Baron of Cameron
(engraving by William Faithorne) 41

The Second Duke of Buckingham
(engraving by Robert White) 43

North Milford, near Tadcaster 45

The Leedes Arabian by John Wootton 46

Flying Childers (attributed to James Seymour) 49

Title page, *A Discourse of Horsemanshippe*
by Gervase Markham, 1593 58

Title page, *The four chiefest Offices belonging
to Horsemanship* by Thomas Blundeville, 1609 59

Title page, *Cavalarice, Or The English Horseman*
by Gervase Markham, 1617 60

English Running-Horse type (etching by Wenceslas Hollar) 66

Four Running-Horse type sprinters near Windsor Castle
(after Francis Barlow) 68

James Stuart, King of Scotland 72

Facsimile of the Minute Book of the City of
Lincoln, England, 1617 73

Sir John Fenwick, "Surveyor of the Race"
at the Royal Stud of Tutbury 74

Coat of Arms of Sir John Fenwick 74

The Duke of Newcastle
(illustration by Abraham van Diepenbeke) 75

Horses of the Duke of Newcastle by Abraham van Diepenbeke 76

Charles I, with the Chevalier de St. Antoine, by Van Dyke 79

Facsimile of the Parliament-ordered Tutbury inventory, 1649 80

Constable Burton, the seat of Sir Marmaduke Wyvill 85

Sedbury Hall, brought to the D'Arcy family
as a part of Isabell Wyvill's dowry 87

James D'Arcy the Younger by Mary Beale 88

Marske by George Stubbs 93

Marske Hall, home of the Hutton family 94

The City and Castle of Aleppo in Syria
(engraving by Drummond) 97

The Darley Arabian (engraving after John Wootton) 99

The Bloody Shouldered Arabian 103

A match race at Newmarket (engraving by Peter Tillemans) 111

The Navesmire Racecourse, York, 1731 112

Monkey 1725 by James Seymour 112

Diomed attributed to George Stubbs .114

Sulieman the Magnificent, Sultan of the
Ottoman Empire by Hans Eworth 123

Midridge Grange in South County Durham,
property of Captain Robert Byerley .128

The Byerley Turk (Fores engraving after John Wootton) 129

The Godolphin Arabian (engraving after George Stubbs
based on a 1753 engraving by John Faber) 132

King's Son of Blank by William Shaw 145

Lexington 1850 by Edward Troye 146

Herod 1758 155

Highflyer by John Boultbee 156

Richard Tattersall by Thomas Beach .157

Eclipse by George Stubbs 161

St. Simon 1881 167

The Tetrarch 1911 by A.G. Haigh 169

Bold Ruler 1954 Pedigree Chart 171

Speed and the Thoroughbred

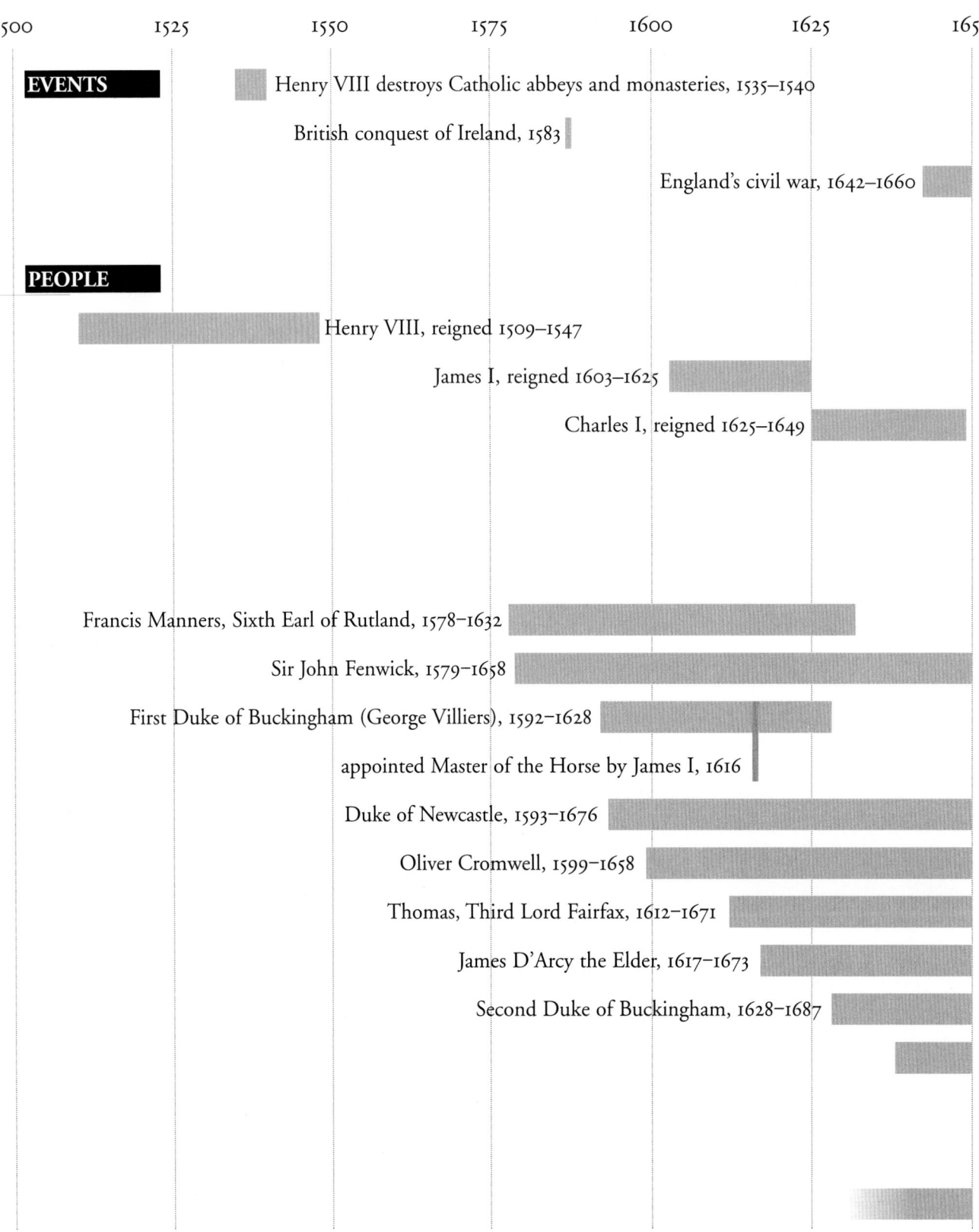

XXII

Timeline

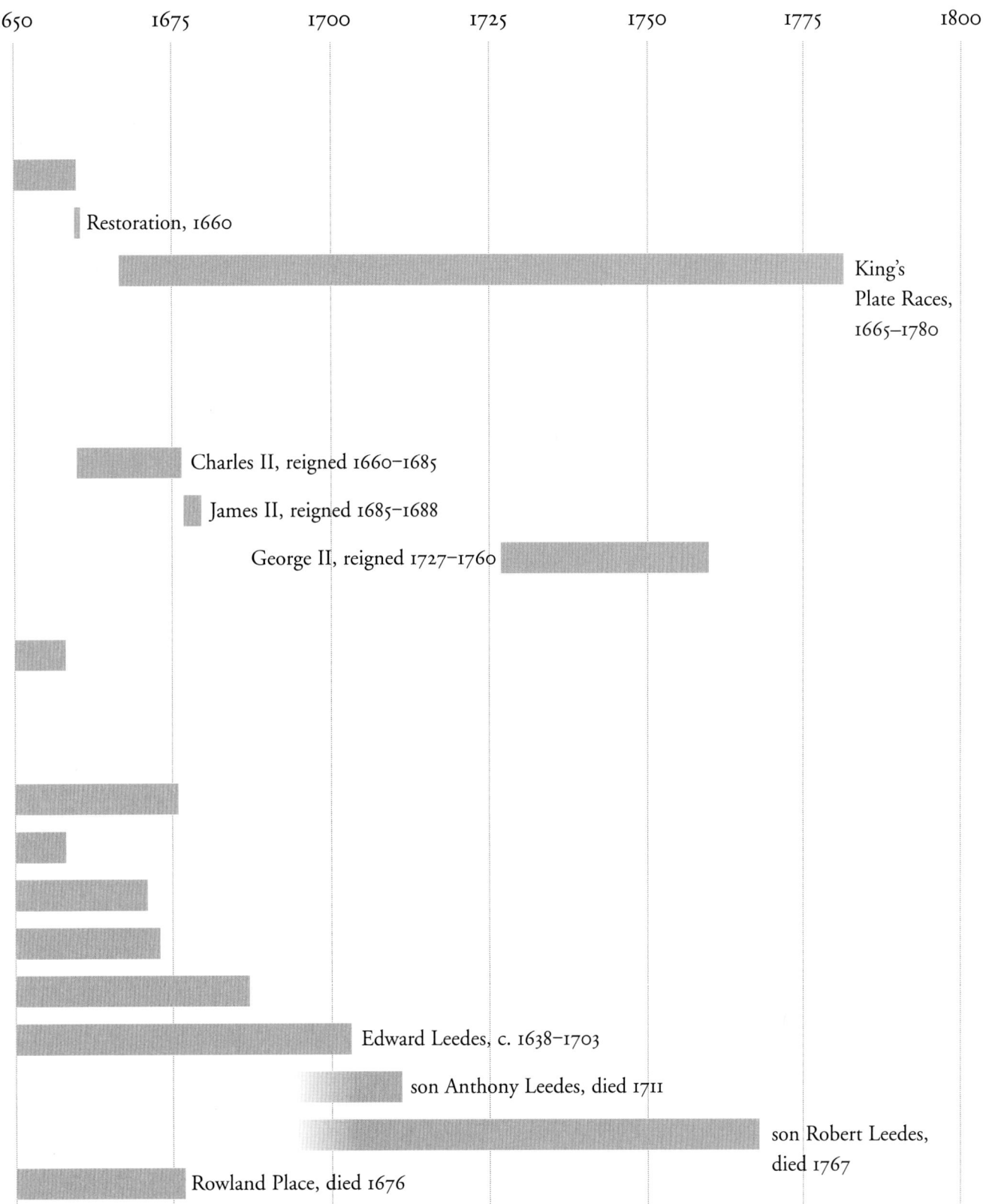

Speed and the Thoroughbred Racehorse

| 1500 | 1525 | 1550 | 1575 | 1600 | 1625 | 1650 |

STUDS

Lady Catherine Manners brings Helmsley Stud to her marriage
to George Villiers, First Duke of Buckingham, 1620
Duchess of Buckingham breeds Old Bald Peg, c. 1635
Third Lord Fairfax captures Helmsley Castle, 1644

Constable Burton (Wyvill Family) before 1640–after 1718
Wellbeck Abbey Stud, Earl of Newcastle forced into exile, 1644
Wallington Stud of Sir John Fenwick
ravaged by Cromwell's soldiers, 1648
Tutbury Royal Stud, 1603–1650
Parliament ordered inventory, 1649
Sedbury Stud (D'Arcy family), c. 1648–1731
Isabel Wyvill brings Sedbury Stud to her marriage to James D'Arcy the Elder, 1648
North Milford Stud (Leedes family), c. 1650–1767

HORSES

Old Bald Peg c. 1635
Old Morocco Barb imported, 1637

Timeline

1650　　　1675　　　1700　　　1725　　　1750　　　1775　　　1800

Helmsley Stud, 1548–1687

Third Lord Fairfax receives Helmsley from Parliament, 1651
Third Lord Fairfax breeds Old Peg (Old Morocco Mare), 1653
Second Duke of Buckingham marries Mary, daughter of Lord Fairfax, 1657
Second Duke of Buckingham breeds Spanker, c.1660–c.1670

Old Peg (Old Morocco Mare) c. 1654
Place's White Turk, imported 1657
Spanker (first Thoroughbred) c. 1670
Byerly Turk at stud, 1691–1702
Old Careless, 1692
Darly Arabian 1700, exported to England, 1704
Betty Leedes c. 1705
Dun Arabian imported to England, 1715
Devonshire (Flying) Childers, 1715–1741
Bartlett's (Bleeding) Childers, 1716
Bloody Shouldered Arabian exported to England, 1719
Godolphin Arabian, 1724–1753
imported, 1730
King Herod, 1758
Eclipse, 1764
Highflyer, 1774

Speed and the Thoroughbred

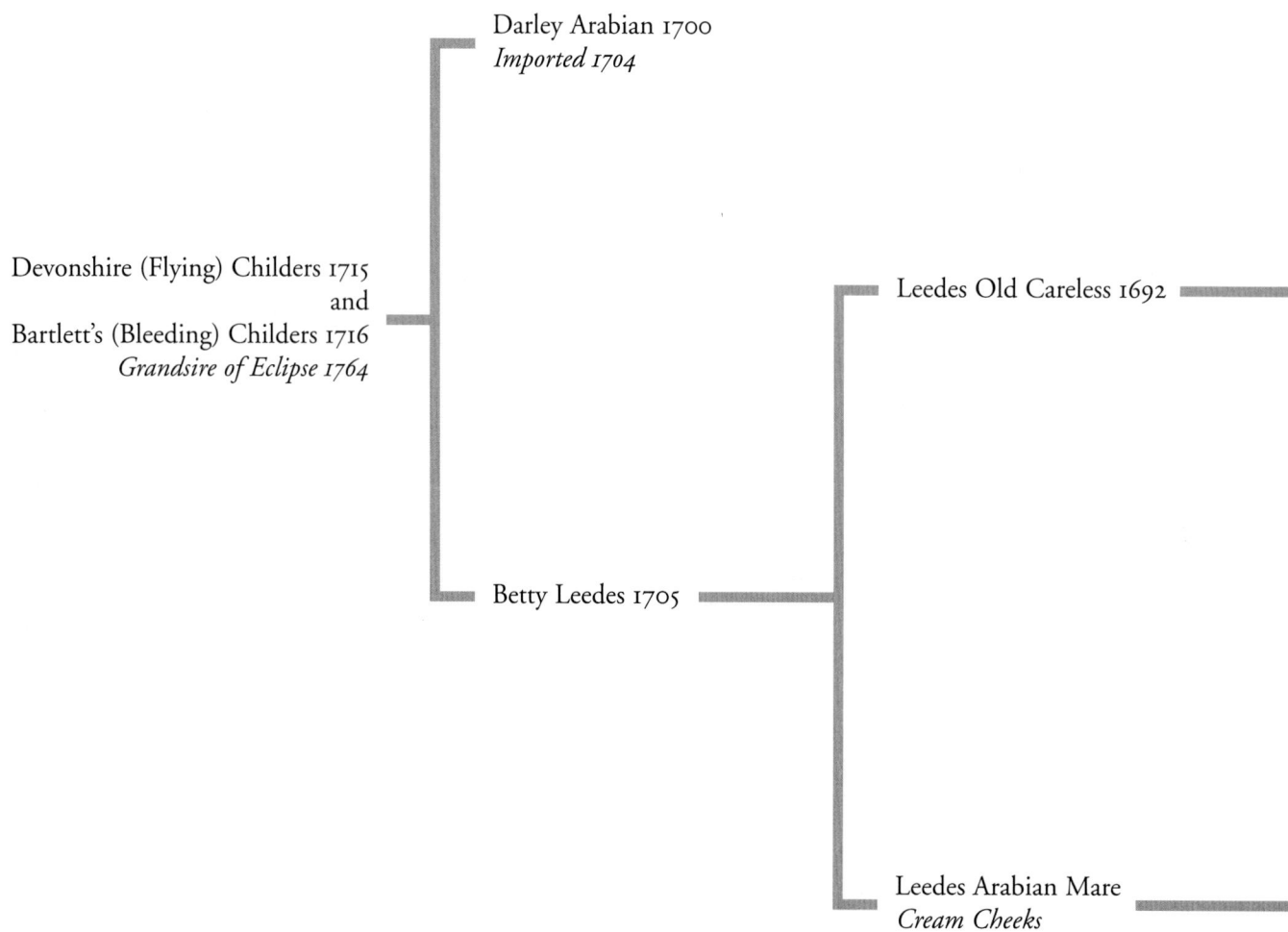

Taproot Pedigree of the Thoroughbred Racehorse

(presumed)
Place's White Turk
imported 1657
D'Arcy Yellow Turk
A Sedbury mare with sprinting speed bloodlines
Spanker
Pelham's Bay Arabian
(presumed)
Old Morocco Barb
Old Peg c.1654
Old Morocco Mare
Old Bald Peg c. 1635
"Barb" Mare
(presumed)
D'Arcy Yellow Turk
Leedes Arabian I
(presumed)
Place's White Turk
imported 1657
D'Arcy Yellow Turk
A Sedbury mare with sprinting speed bloodlines
Spanker
Pelham's Bay Arabian
(presumed)
Old Morocco Barb
Old Peg c.1654
Old Morocco Mare
Spanker Mare
North Milford Stud
Old Bald Peg c. 1635
Old Morocco Barb
Old Peg c.1654
Old Morocco Mare
Old Bald Peg c. 1635

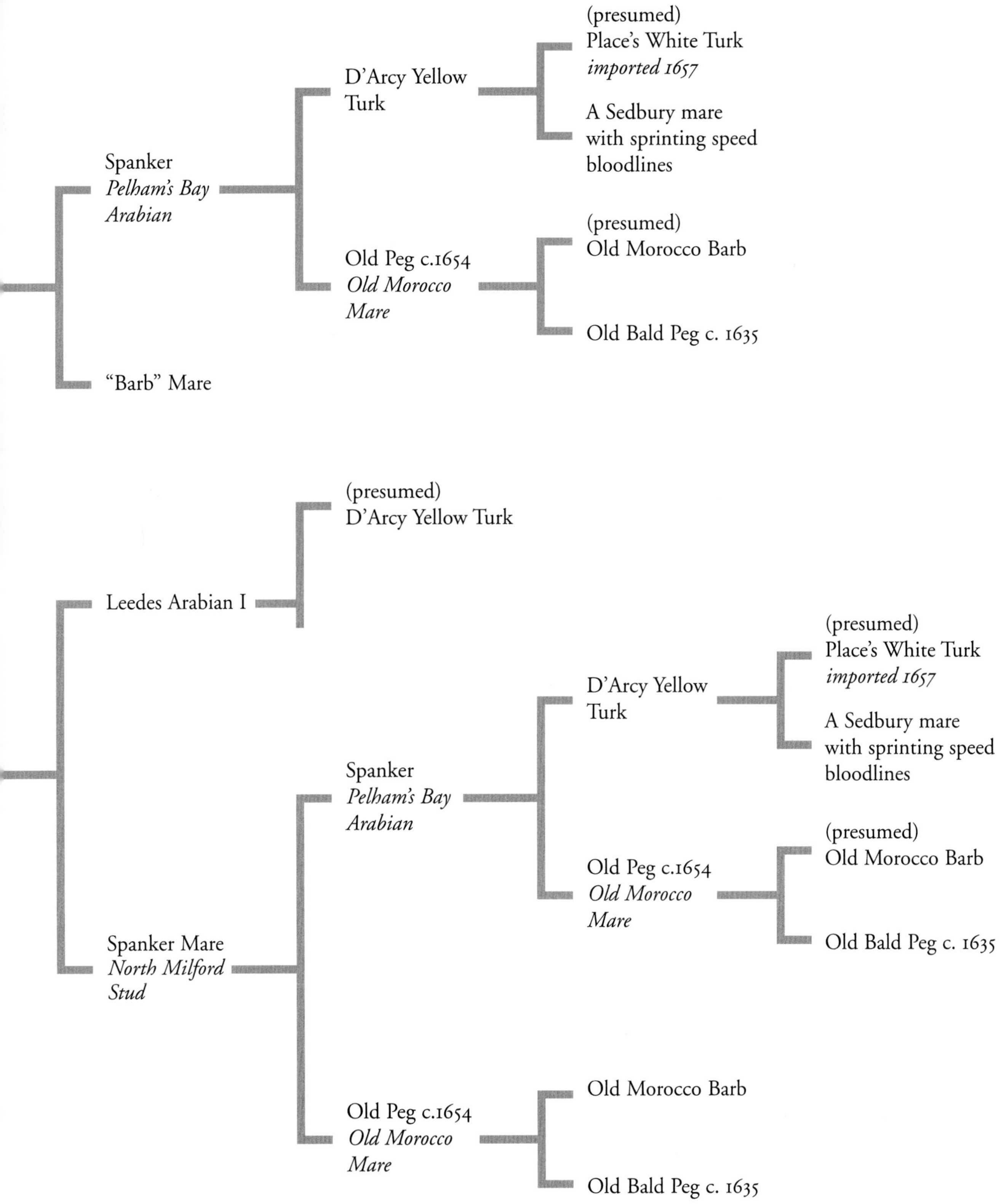

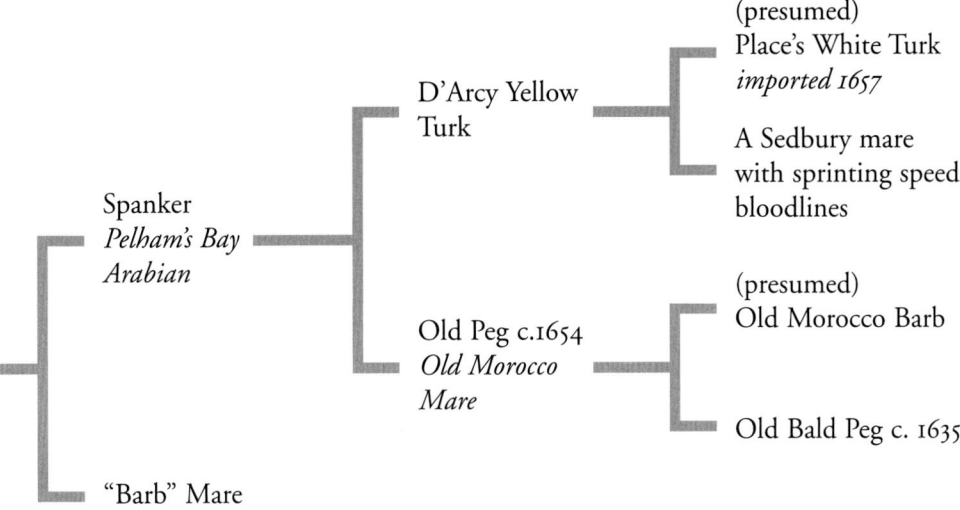

CHAPTER I

An Overview

Sources of Speed

Speed is the bottom line of horse racing and racehorse breeding. For the past 500 years and more, the British racehorse has been the world's fastest horse. Since the mid-eighteenth century it has been known as the Thoroughbred.

Thoroughbred bloodline speed strains are found in the pedigrees of horses registered in Volume I of the *General Stud Book* (*G.S.B.*). The most recent edition of Volume I (1891) is the reference book for *Speed and the Thoroughbred*. Only when connected with these pedigrees can other turf records be authenticated as Thoroughbred history. By making these connections, *Speed and the Thoroughbred* identifies two native sprinting speed strains and one imported middle distance speed strain. The combination of these three strains in its foundation is the source of Thoroughbred speed. This is the first comprehensive history of Thoroughbred breeding and bloodlines to be based on these three speed strains.

The two native short course (approximately a quarter of a mile) sprinting speed strains were those of the pre-Christian Irish Hobby and the sixteenth century English Running-Horse. The third strain of Thoroughbred speed came from quite another source. Because export of purebred Arabians was forbidden, the Turkish Sultans provided a special strain of diplomatic gift Turcoman stallions, their beauty enhanced by pedigree Arabian crosses. Two of these Turcoman-Arabian imported stallions provided the middle distance (4-mile course) speed required by Charles II's King's Plate races (1665). The two, Place's White Turk and the Godolphin Arabian, imported in 1657 and 1730 respectively, provided the third source of Thoroughbred speed. They appear in the ancestry of virtually all the foundation mares of the Thoroughbred breed.

Hobbies: The First Source of Speed

Before the Restoration of Charles II (1660), most racing in the British Isles was organized to attract large crowds of buyers—in Ireland at the great fairs and in England (c. 1512–1642) at annual town races. So that everyone could see every horse from start to finish, races were for sprinters over short courses.

The earliest Thoroughbred sprinting speed strain, known as the Hobby, traces back to Irish fairs in pre-Christian times. There was racing at the Curragh of Kildare (still the center of Irish racing) in King Conari's reign (first century A.D). The speed of this bloodline is unique, not achieved by any other country. Preserved by *female* lines, it is this strain which continues to make the Thoroughbred the world's fastest horse. Irish Hobbies were imported into England by Henry VIII, by Yorkshire breeders (1528) and by James I (reigned 1603–1625).

The Helmsley Stud

The Earls of Rutland, who then and now reside at Belvoir Castle, bred Hobby strain mares at their North Yorkshire Helmsley stud in the sixteenth and seventeenth centuries (c. 1548–1687). On February 21, 1596, the Fifth Earl's Hobby won a race and a bet of £45 at the Forest of Galtres course near York.

About 1635 at Helmsley, the Fifth Earl's niece, wife of the First Duke of Buckingham, bred Old Bald Peg, the most important Hobby strain Thoroughbred foundation mare. Old Bald Peg appears in the extended pedigree of virtually every present-day Thoroughbred.

The Old Morocco Mare

In 1644, Oliver Cromwell's famous cavalry General, the Third Lord Fairfax, besieged, captured and razed Helmsley Castle, driving the young Second Duke of Buckingham and his mother into exile. In 1651, Parliament gave Fairfax the entire Helmsley estate and its livestock as "a salve for an old wound" received during the siege of Helmsley. Fortunately, about 1653, Fairfax bred Old Bald Peg to the Old Morocco Barb, probably one of four stallions presented in 1637 to Charles I by the Emperor of Morocco. The outcome was Old Peg, known more commonly as the Old Morocco Mare. It is this *General Stud Book* connection (p.14) which establishes the pre-Christian Irish Hobby as a prime source of Thoroughbred sprinting speed.

In 1657, the Second Duke of Buckingham, then 29 years old, secretly stole back from exile to the Fairfax residence, Nun Appleton, and married Mary, the General's heir and only child, thereby recovering his inheritance, Helmsley and its livestock. When the Old Morocco Mare was bred by Buckingham to the D'Arcy Yellow Turk, she produced Spanker whom the *General Stud Book* calls "the best horse at Newmarket in Charles II's reign" (1660–1685): the best because he was the leading winner of 4-mile multiple heat King's Plates. Spanker was also the leading British-bred sire of the seventeenth century, the first registered horse whose pedigree contained all three speed strains, and therefore the first Thoroughbred.

Running-Horses: The Second Source of Speed

The second source of Thoroughbred sprinting speed is the Running-Horse, described in detail by Gervase Markham in his 1593 *Discource of Horsemanshippe* (reprinted 1594–1617, in several editions). This is said to be the first book on training racehorses. Running-Horses were sprinters who competed in annual sprint races over short courses, which on market days attracted large crowds to the several towns and cities which organized them c. 1512–1642. The crowds attended churches and patronized merchants, inns and tavern keepers.

The Tutbury Royal Stud (1603–1650)

When James Stuart, King of Scotland, became James I, King of England (reigned 1603–1625), his personal property included the Tutbury Royal stud and its band of Running-Horse brood mares, to which he added Irish Hobbies. A letter written to the Earl of Northumberland in 1610 mentions "the maire that Sir John Fenwick gave to the King that was the swiftest horse held to be in England." Fenwick was England's largest breeder of sprinting Running-Horses.

James I loved sprint racing. The race which entertained the King on April 3, 1617, is recorded in the Note Book of the city of Lincoln:

> On Thursday there was a great horse race on the heath for a cupp where his MA^{tie} was present and stood on a scaffold the citie had caused to be set up, and with all caused the race a quarter of a mile long to be railed and corded with ropes and stoops on both sides, whereby the people were kept out and the horses wch ronned [which ran] were seen faire.

The Duke of Newcastle (1593–1676), famous for his 1657 Antwerp book on horsemanship (*Nouvelle Methode et Invention Extraordinaire de Dresser Les Chevaux*), was also a breeder of Running-Horses. In his *A New Method and Extraordinary Invention to Dress Horses* (London, 1667, p. 80), he wrote:

> I like the sport [of] a Running-Horse very well. I think myself as Good a *jockey* as any, and have Ridden many Hundreds of *Matches*.

Newcastle, a trusted adviser of Charles I (reigned 1625–1649), helped to secure the appointment of Sir John Fenwick (1579–1658) as Surveyor of the Tutbury stud.

Following the execution (January 30, 1649) of Charles I by Parliament under Oliver Cromwell, the Tutbury horses were sequestered and a committee was appointed by Parliament to make an inventory. On July 24, 1649, the committee made their inspection and named the mares according to their breeding. This inventory provides the only information available about the origin and breeding of the Tutbury horses. After the inventory was completed, the Running-Horse and Hobby strain mares of the Tutbury Royal stud were sold or given away indiscriminately.

Turcoman-Arabians: The Third Source of Speed

The Merry Monarch's King's Plates 1665

In 1660, two years after Cromwell's death, the monarchy was restored and Charles II (1630–1685) succeeded to the throne. Aware of the deficiency of his late father's cavalry, Charles II prepared "Articles" for a new series of races which called for two to four heats of 4 miles each with 12 stone (168 pounds) rider weights. This instantly necessitated the addition of middle distance speed strains to the sprinting bloodline mainstream, and abruptly changed the evolution of British racing. On October 16, 1665 at Newmarket, Charles II presented the first of these awards, called the King's Plates. Over 20 Plates were awarded annually at various racing centers throughout the eighteenth century, as reported in the *Racing Calendars* (1727, ff).

Rowland Place and the White Turk 1657

Rowland Place, who had left his Royalist family to serve as Oliver Cromwell's Horse Master, provided the key link to the third source of Thoroughbred speed. In November 1657, instead of a requested Arabian stallion, there arrived at the Hampton Court stud a light grey, later white, Turk (Turcoman) stallion, a diplomatic gift of Sultan Mohammed IV to Cromwell. This stallion was entrusted to the care of Rowland Place. Cromwell died September 3, 1658. The Parliamentary government was disintegrating. Place had become deeply attached to the magnificent white stallion, who was then threatened with obscurity unless bred to good mares. There was then no functioning authority from which to ask permission, but Place nonetheless moved the horse to Dinsdale, his family estate in the southern part of county Durham, adjacent to the Running-Horse and Hobby studs of North Yorkshire, which contained the best sprinting strain mares in England. Bred to these mares, the White Turk became the most influential sire of the seventeenth century, contributing the middle distance speed needed to win King's Plates (begun 1665). Place's White Turk is registered in the *G.S.B.*, p. 388.

Although he had thus achieved the stallion's success, Place must have realized this move would ruin the rest of his career (he lived until 1676). Place was already hated as a traitor by the Roman Catholic Royalists. He was now hated as a thief by the Protestant Roundheads, the followers of Cromwell. Because of political hatreds, the stud career of Place's White Turk was virtually excluded by John Cheny in his 1743 *Racing Calendar*, and was limited to faint praise. He was listed only as the sire of the little-known Commoner and Wormwood. This was copied verbatim in the 1791 *General Stud Book*, Volume I and subsequent editions (1793–1891). Purposely omitted, the evidence suggests, was the ancestry of the White Turk's best sons and grandsons, the publication of whose pedigrees would have revealed and honored their tail male descent from him. These were the D'Arcy Yellow Turk and D'Arcy White Turk, the Second Duke of Buckingham's Helmsley Turk and the Byerley Turk (not imported, C. M. Prior, *Early Records of the Thoroughbred Horse*, 1924, p. 143).

Publicly, the owners of the principal studs, including Lord D'Arcy, avoided mentioning the name of Rowland Place, even when writing a pedigree (Prior, *Early Records of the Thoroughbred Horse*, p. 36). Privately, they bred their best mares to Place's stallion. The White Turk provided the middle distance speed needed to win 4-mile, multiple heat King's Plate races begun in 1665. This was the third and final source of Thoroughbred speed.

The three sources of Thoroughbred speed were finally combined: the sprinting speed strains of the Hobbies and the Running-Horses, and the middle distance speed strain of one unique horse, Place's White Turk. It was not until 1730, sixty-three years later, that a fresh outcross, the gift of another Turkish Sultan, reached England to augment the middle distance (third) speed strain of the Thoroughbred. This was the Godolphin Arabian.

The Godolphin Arabian 1724–1753

The Godolphin Arabian was obtained (c. 1730) through diplomatic channels by the Duke of Lorraine, through his alliance with the Austrian royal family. The prepotence of this stallion was miraculous. Singly, he passed on his part-Turcoman and part-Arabian bloodlines, his conformation, his temperament and his middle distance speed genes, not only to his immediate descendants but also to the entire Thoroughbred breed.

Like Place's White Turk more than a half-century earlier, the Godolphin Arabian contributed Turcoman-Arabian middle distance speed bloodlines—the third and final source of Thoroughbred speed. The Godolphin's deep and sloping shoulders, passed on to his descendants, enabled greater extension and greater speed at the gallop. The veterinary surgeon, William Osmer, in his 1756 *Dissertation on Horses*, wrote:

> there never was a horse so well entitled to get racers as the Godolphin Arabian: for, whoever has seen this horse must remember, that his shoulders were deeper and lay further into his back, than any horse ever yet seen. Behind the shoulders, there was but a very small space ere the muscles of his loins rose exceeding high, broad and expanded, which were inserted into his quarters with greater strength and power than in any horse, I believe, ever yet seen.

It is these shoulders and quarters which distinguish the Thoroughbred from other breeds. The Godolphin Arabian was the foundation sire of the twentieth century Thoroughbred.

Threats to the Thoroughbred Breed: Impossible Problems—Miraculous Solutions

During the seventeenth century, the future of the young Thoroughbred breed was threatened by seemingly impossible problems for which miraculous solutions were found. Today, Thoroughbred bloodlines are in the pedigrees of most of the world's registered athletic horse breeds, and most of the unregistered sport horses have Thoroughbred crosses in their pedigrees. Part-Thoroughbreds make up the world's largest, most widely dispersed group of pedigreed horses and ponies. How did the young breed of racehorse survive the threats to its development and flourish?

Irish Hobby Exports

Because of excessive exports, constant local warfare and the British conquest of Ireland (1583), the Hobby breed was almost extinct in Ireland by the year 1600.

Running-Horse

The next sprinting speed strain to disappear from English racing was the Running-Horse. On October 16, 1665, when Charles II founded the series of race awards known as King's Plates, his "Articles" called for higher weights of 12 stone (168 pounds) and longer distances (two to four heats of 4 miles each over round courses). The era of sprint racing (quarter of a mile) in England was coming to a close and so was the Running-Horse breed. Its bloodlines, fortunately, have been carried on by the *female* lines of the Thoroughbred breed.

Small Regional Breed

During the second half of the seventeenth century, the Thoroughbred was a small regional, North Yorkshire breed with a narrow foundation of Hobby and Running-Horse strain bloodlines. By moving the center of racing from York to Newmarket following his Restoration in 1660, Charles II inadvertently but fortuitously contributed to enlarging the geographic marketplace for the young Thoroughbred breed.

Civil War

The foundation mares of the Thoroughbred breed were nearly lost during England's civil wars (1642–1660). The major stud farms were owned by Roman Catholic "Royalists," followers of Charles I, and were hated by the Protestant "Roundheads," followers of Cromwell. In 1644, the Earl of Newcastle, hero of many battles, was forced into exile, and left behind the Running-Horse mares of his Welbeck Abbey stud. In 1648, Sir John Fenwick's Wallington stud, the largest group of Running-Horse mares in Britain, was ravaged and its mares scattered by marauding soldiers.

Following the execution of Charles I by Parliament under Oliver Cromwell, the horses of the Tutbury Royal stud were sequestered. On July 24, 1649 a committee appointed by Parliament made an inventory, naming the mares according to their breeding. The Tutbury Running-Horse and Hobby strain mares were then sold or given away indiscriminately. The young Thoroughbred breed seemed to be doomed.

Lord Fairfax at Helmsley 1651–1657

Having been given the North Yorkshire Helmsley stud and its Hobby strain mares by Parliament (1651), it was the Third Lord Fairfax who began the recovery of the Thoroughbred. By breeding the Old Morocco Barb to the famous Hobby strain foundation mare Old Bald Peg (born c. 1635), Fairfax supplied the breed with one of its most important tail female lines.

The Sedbury Stud of the D'Arcys

Mares with Running-Horse and Hobby bloodlines, scattered during the English civil war, were salvaged by James D'Arcy the Elder (1617–1673), the breed's second rescuer. His wife, Isabel Wyvill, contributed the North Yorkshire Sedbury stud as part of her dowry. The only clues to the identity of the names are those from the sparse summer 1649 inventory of the Tutbury Royal stud. (C. M. Prior, *Royal Studs*, 1935, p. 56) The mares carrying the name Fenwick or Newcastle indicate Running-Horse bloodlines, acquired from Sir John Fenwick and the Earl of Newcastle. The mares named Carleton indicate Hobby bloodlines, acquired from Thomas Carleton. During the 1650s, D'Arcy collected some of the Tutbury mares of Charles I that had been dispersed after the Parliament-ordered 1649 inventory. Several of these and their descendants are registered as Royal Mares in the *General Stud Book*. He also collected the scattered mares from Fenwick's Wallington stud.

The North Milford Stud

Principal market breeders of Thoroughbreds in the seventeenth century were Edward Leedes (c. 1638–1703) of North Milford stud near Tadcaster, not far south of York, who was succeeded by his son Anthony (died 1711). Their foundation mares, the second and fourth foals of the Old Morocco Mare, were Young Bald Peg and the Spanker Mare (*G.S.B.*, pp. 14, 17). By selling the beautifully bred descendants of these mares to stud farms in different sections of the country, the Leedes helped to expand the North Yorkshire local strains into a national breed.

The Darley Arabian 1700

Sometime after 1516, the Ottoman Turkish sultans forbade the export of the highly prized pure Arabian horses. No Arabians imported into England during the seventeenth century are registered in the General Stud Book. (The Markham Arabian, purchased by James I in 1616, was probably British-bred, as discussed in Chapter VII.) In the eighteenth century, three Arabian stallions were smuggled out of Turkey and registered in the General Stud Book: the Darley, the Oxford Dun and the Bloody Shouldered Arabians.

The export of the Darley Arabian (*G.S.B.*, p. 391) in 1704 was achieved by the Aleppo merchant Thomas Darley through his membership in an Aleppo hunting club. In 1715, when another English Aleppo merchant, Nathaniel Harley, managed to send the Dun Arabian (*G.S.B.*, p. 389) to his nephew, Lord Oxford, he described his difficulties in a letter dated February 15th:

> Three Expresses have been sent after him [the stallion] and all the passes of the Mountains between this [Aleppo] and Scanderoon ordered to be watched, and ye marine strictly guarded to prevent his being ship'd off. (Prior, *Early Records of the Thoroughbred Horse*, 1924, p. 141)

Harley also succeeded in shipping the Bloody Shouldered Arabian (*G.S.B.*, p. 391) to Lord Oxford in the winter of 1719–1720, the last of the three imported pure Arabian sires.

In 1714 and 1715, the Darley Arabian was bred to Leonard Childers' mare, Betty Leedes (c. 1705). By Old Careless 1692, a son of Spanker, this mare was a granddaughter of the Spanker Mare, referred to previously, bred at North Milford by Anthony Leedes. Betty Leedes' pedigree contained all three strains of the sources of Thoroughbred speed, including three crosses of the Helmsley Hobby strain mare Old Bald Peg. The first mating produced Devonshire (Flying) Childers 1715, "the fleetest horse ever trained in this or any other country" (*G.S.B.*, p. 379). Although the

General Stud Book called the horse Devonshire Childers, he was more commonly referred to as Flying Childers. Purchased in 1719 by the successful breeder, the Second Duke of Devonshire, he was retired to the latter's Chatsworth stud farm beside the magnificent Chatsworth mansion and its world famous library. Flying Childers' extraordinary speed made him and his brother, Bartlett's (Bleeding) Childers 1716, the two most popular stallions of the 1720s. Through these two full brothers, the Darley Arabian spread his high spirits, endurance, stamina and classic beauty, *but not speed*, throughout the Thoroughbred breed. Over ninety percent of all Thoroughbreds today descend tail male from the Darley Arabian 1700 through Bartlett's (Bleeding) Childers 1716 and Eclipse 1764.

Imported Morocco Barbs

Charles II was one of several monarchs who received gifts of beautiful stallions, an important part of mid-Eastern diplomacy. The Emperors of Morocco, North Africa, presented to Charles I, to Charles II and to Louis XIV of France, seven "Barbs," registered in the *General Stud Book*, imported in 1637 (the Old Morocco Barb), c. 1675, and 1698. Documentary evidence that the Emperors purchased many of these gift stallions, who were part Arabian, mostly in Egypt, is cited by Lady Wentworth (*Thoroughbred Racing Stock*, 1938, pp. 235–236).

Turcoman-Arabian Sires

Speed developed both in desert warfare and more than a thousand years of tribal racing made Arabians the world's fastest horses over long distances. British breeders had no personal knowledge of these Arabians, and even the widely traveled Duke of Newcastle had only seen "but one right Arabian" (*A New Method and Extraordinary Invention to Dress Horses,* London, 1667, p. 63). Nevertheless, British breeders believed Arabians would sire middle distance speed King's Plate winners.

Winchelsea

About 1663, Charles II sought to acquire an Arabian stallion through diplomatic channels from the Turkish Sultan, but his request was ignored. Undeterred, he urged his Ambassador to Turkey, Lord Winchelsea, to keep trying. From Pera, on February 24, 1666, the ambassador wrote he

An Overview

Flying Childers by John Wootton (1682–1765), c. 1720.
The General Stud Book *(p.14) calls Flying Childers (by the Darley Arabian out of Betty Leedes) "the fleetest horse ever trained at Newmarket or in any other country." Foaled in 1715 for Leonard Childers of Doncaster, he raced for the Duke of Devonshire and won every race he started.*

The Wootton painting depicts the aftermath of his most famous race. In 1721, Flying Childers, carrying 9 stone 2 pounds, ran against his half-brother, Almanzor (by the Darley Arabian), and the Duke of Rutland's Brown Betty (by Basto), each carrying 8 stone 2 pounds. The race was over the round course at Newmarket: 3 miles 6 furlongs and 93 yards. Flying Childers' time was 6 minutes 40 seconds which equates to moving at a speed of 50 feet per second, close to 35 miles per hour. The painting shows Flying Childers, the bay, being rubbed down after this run. Almanzor is the gray, and Brown Betty, still saddled, stands on the left.

was going to "Byk Be Zar, 13 days journey hence... to try for some of the Turcoman breed." Native to the region between the Black and Caspian Seas, Turcoman bloodlines provided middle distance speed needed to win King's Plates.

Since gifts of beautiful stallions were an integral part of mid-Eastern diplomacy, and since the export of pure Arabians was forbidden, Lord Winchelsea's alternative suggestion "to try for some of the Turcoman breed" had previously been adopted.

Arabian Crosses in the Turcoman Horses

Place's White Turk

Place's White Turk was a Turcoman stallion with Arabian crosses added for high spirits and greater beauty. The *General Stud Book* (p. 388) notes the then common practice of calling horses Arabians if they had a horse of that breed in their pedigree. The evidence that the pedigree of Place's White Turk contained an Arabian cross is provided by the *General Stud Book* registrations of the Pelham's Bay Arabian and the Oglethorpe Arabian.

After the death of the Second Duke of Buckingham in 1687, his Helmsley Spanker, the best racehorse and sire of the seventeenth century, was acquired by Charles Pelham of Brocklesby Park, a prominent breeder, who renamed the horse Pelham's Bay Arabian. The Arabian blood in the pedigree of Spanker came through his sire, the D'Arcy Yellow Turk, from his grandsire, concluded to be Place's White Turk, imported in 1657. (See Chapter IX.) Another stallion by the D'Arcy Yellow Turk, known as the Oglethorpe Arabian (*G.S.B.*, p. 389, no pedigree), was the property of Sutton Oglethorpe who became Master of the Studs to Charles II on the death of the previous Master, James D'Arcy the Elder (d. 1673).

The Godolphin Arabian

The stallion that came to be known as the Godolphin Arabian was sold by the Duke of Lorraine in 1729 to Edward Coke, Longford Hall, Derbyshire, who imported him into England the following year. By calling his young Turcoman stallion an Arabian, Coke was able to charge a higher stud fee. Had he advertised his young import as a Turcoman stallion, he could not have charged a stud fee of more than one guinea.

The May 10, 1746 issue of *The Mercury* newspaper, published in Stamford, Leicestershire, illustrates the low stud fee for Turcomans. An advertisement for an unnamed stallion, placed by the famous livestock improvement breeder Robert Bakewell of Dishley farm, reads:

> A brown-bay stallion of the Turcoman breed, very strong, bought out of the Grand Signor's [Sultan of Turkey's] stud. One Guinea the mare and two shillings and sixpence the servant.

In spite of the horse's royal origin and Bakewell's reputation, because the Turcoman breed was little regarded, the stallion's stud fee was only what was commonly charged for breeding to farm mares.

H.R.H. the Duke of Cumberland

H.R.H. the Duke of Cumberland (1721–1765), second surviving son of George II, had a profound knowledge of seventeenth and eighteenth century pedigrees, which enabled him to breed two (of the three) Great Progenitors, namely King Herod 1758 and Eclipse 1764. Cumberland knew the Godolphin was not an Arabian. He ordered the 1753 portrait of the Godolphin, painted from life by his household artist David Morier. On a 1753 print from this portrait, the Duke's inscription reads: "This extraordinary foreign horse got a greater number of fine horses with just temper and superior speed than ever any Arab did." By referring to the horse as "foreign" and not as an Arab, the Duke makes the distinction between the Godolphin and a pure-bred Arabian. Cumberland understood the Godolphin Arabian and Place's White Turk, imported 73 years earlier, were both diplomatic gift Turcoman-Arabian stallions with middle distance (4-mile course, multiple heat) speed bloodlines.

King Herod 1758

Events suggest that the Duke was also convinced that the Byerley Turk, at stud 1691–1702, was a tail male descendant of Place's White Turk. In spite of this latter stallion's obscurity caused by religious and political hatreds, and although the tail male line of Place's White Turk was 100 years old in 1757, Cumberland believed he could still produce a first class racehorse, without using Godolphin bloodlines. He accomplished this by breeding his mare, Cypron, to Tartar 1743, registered tail male descendant of the Byerley Turk.

The result was the first-class racehorse King Herod 1758. In his 1803 *Turf Register* (p. 297), William Pick called him "one of the best-bred horses this kingdom ever produced—sire of a larger number of racehorses, stallions and brood mares than any other horse, before or since." This tail male line extends from the White Turk in 1657 through King Herod and Highflyer 1774 to The Tetrarch 1911, sire of the famous matron Mumtaz Mahal 1921.

Eclipse 1764

In 1763, in order to convince British breeders planning matings that they were paying too much attention to the racing records of sire and dam, and too little attention to the rest of the pedigree, the Duke of Cumberland went no farther than his own paddocks near Cumberland Lodge in Windsor Great Park. For a sire, he chose the 13 year old Marske whose unsatisfactory racing record had been rewarded with farmers' mares at a guinea apiece. Marske was bred by John Hutton III, grandson of Lord D'Arcy of Sedbury. For a dam Cumberland chose the 14 year old unraced Spiletta (by the Godolphin Arabian) whose only foal had never raced. The rest of the six-generation pedigree of Marske's and Spiletta's chestnut colt with a white leg was filled with the best seventeenth century Hobby and Running-Horse sprinting speed bloodlines.

It was a tragedy that the Duke died October 31, 1765. In the dispersal auction conducted by Richard Tattersall the following year, the chestnut colt, with no asset except his conformation, brought only 75 guineas, the bid of William Wildman, a Smithfield market grazier and meat merchant. Because the colt was born April 1, 1764, when the moon almost covered the sun, he was named Eclipse.

As a 5-year-old, Eclipse made his first start in a five horse race at Epsom on May 3, 1769. The ease with which he won the first heat so astounded an Irish professional gambler, Dennis O'Kelly, that O'Kelly bet all comers at long odds he could name the official order of finish in the next heat. He won a fortune when, as he had written, the stewards posted "Eclipse first and the rest nowhere"—their description when a winner distanced the rest of the runners by 200 yards or more.

The combination of Eclipse's careers as an unbeaten racehorse and as a sire was unsurpassed during the eighteenth century. His prepotence was a major factor in extending the tail male line of the Darley Arabian 1700, the only tail male line of Thoroughbred racehorses which survives unbroken since the eighteenth century to this day.

An Overview

Eclipse with Mr. Wildman and His Sons
by George Stubbs (1724–1806), oil on canvas.
Eclipse (by Marske out of Spiletta), the most famous and influential horse of the eighteenth century, was the outcome of the Duke of Cumberland's belief that British breeders were paying too much attention to the racing careers of sires and dams and not enough attention to pedigree bloodlines. Marske's racing record was disappointing and neither Spiletta (by the Godolphin Arabian) nor her only foal had ever raced, but their pedigrees were filled with the best seventeenth century Hobby and Running-Horse sprinting speed bloodlines. Unfortunately, the Duke did not live to enjoy the triumph of this mating and the vindication of his breeding theories.

Sprinting Speed in the Nineteenth and Twentieth Centuries

The Thoroughbred breed was established. However, in order to be maintained, middle distance and other longer racing speeds must be reinvigorated periodically by sprinting speed bloodlines.

Following excessive use of middle distance mile-and-a-half Derby and Oaks winner strains during the nineteenth century, the breed needed 'refreshing.' One of the earliest breeders to act accordingly was Viscount Falmouth of the Mereworth Castle stud (A. S. Hewitt, *The Great Breeders and Their Methods*, 1982). In 22 years, Falmouth bred nineteen winners of the five English classic races for 3-year-olds (Derby, Oaks, etc.). Of the twenty-four brood mares at Mereworth in 1880, twenty-two had won races as 2-year-olds, most of these at sprinting distances.

Over the next 35 years, six stallions surfaced who were especially dominant for sprinting speed:

- St. Simon 1881, bred and owned by the Sixth Duke of Portland;
- Domino 1891, bred by Barak Thomas, acquired by James R. and Foxhall Keene, the Castleton stud, Lexington, Kentucky;
- Americus 1892, bred in California by "Lucky" Baldwin, acquired by Richard ("Boss") Croker of Tammany Hall, New York, who took him to Ireland;
- The Tetrarch 1911, owned by Atty Persse, famous Irish trainer of sprinters, the multiple winner of sprint races, tail male descendant of Place's White Turk 1657;
- Phalaris 1913, a top-class sprinter who the Sixteenth Earl of Derby fortunately failed to sell for £5000 during World War I; and
- Havresac II 1915, to whom the famous Italian breeder, Federico Tesio, sent his mares; sire of Nearco 1935 and grandsire of Ribot 1952.

All six appear in the pedigree of Bold Ruler 1954 and in the pedigrees of other notable twentieth century Thoroughbreds.

Speed and the Female Bloodlines

As will be seen, speed is transmitted through *female* lines. This has been demonstrated time and again. The premier seventeenth and eighteenth century English breeders seem to have understood this. They cherished, protected and ofttimes rescued their sprinting strain brood mares. The concept of speed deriving from female lines is plainly evidenced by the fact that the great majority of Thoroughbreds living today trace their ancestry back through more than fifteen unbroken tail female lines, but only one thriving tail male line.

Foundation Mares

The first nineteen pages of the 1891 *General Stud Book* list 78 of the earliest (late seventeenth century) foundation mares. On page 6 of the introduction to his *Early Records of the Thoroughbred Horse*, 1924, C. M. Prior comments:

> It has hitherto escaped notice that these seventy-eight mares were domiciled in [North] Yorkshire of which Bedale was the center point...the exceptions are but four or five...The [Running-Horse] studs, those of James, Lord D'Arcy of Sedbury and [the Hobby strain stud] of the Second Duke of Buckingham at Helmsley, [North] Yorkshire...were originally responsible for most of these foundation mares. (Sedbury stud, c.1648–1731; Buckingham, 1628–1687)

In 1895, the Australian turf scholar C. Bruce Lowe (*Breeding Racehorses by the Figure System*) identified the earliest 43 late seventeenth century tail female performance families, numbered in accordance with the total wins by their descendants in the five English classic races for 3-year-olds. Of special importance are mares numbered *1–15*, which have sprinting speed backgrounds. Of the 64 stallions and mares in the sixth generation of the pedigree of Bold Ruler 1954, 63 trace back *in tail female* to one of these first fifteen late seventeenth century tap root mares. The successes of many present-day breeders are explained by the pedigrees of their brood mares. Many of these pedigrees trace back to Bruce Lowe's first fifteen in tail female lines.

Bruce Lowe's number 5 was Mr. Massey's Black Barb Mare, and 6 was Old Bald Peg c. 1635. Both were Hobby strain mares, the property of the Rutland family's Helmsley stud in North Yorkshire. The families numbered 2, 3, 4, 7, 8, 11, 12, 13 and 15 were the property of the North Yorkshire Sedbury stud (c. 1648–1731) of James D'Arcy, father & son. These were the foundation mares registered in the *General Stud Book* who had Running-Horse and Hobby backgrounds, five of whom plus their female descendants were registered in the *Stud Book* as Royal Mares numbers 7, 11, 12, 13 and 15. These families descend from mares in the Tutbury stud of Charles I who reigned from 1625 to 1649, and the Wallington stud of Sir John Fenwick which was dispersed in 1648.

Seventeenth Century to Today

We thus have approximately 15 tail female lines, most of which, for the past 300 years, have dominated Thoroughbred breeding and which have survived unbroken. Only one tail male line of racing Thoroughbreds survives (Darley Arabian imported 1704). This was not a middle distance speed line, however. Authentic Arabians have no middle distance speed. The speed in the Thoroughbred comes from the mares to which the succession of sires were bred. This 15:1 ratio affirms the beliefs of the Arabian Bedouins and other tribal breeders, who based their successful breeding practices on female lines since long before their beautiful horses first enchanted the invading armor-clad horsemen of the Crusades.

CHAPTER II

THE IRISH HOBBY

Early Irish Racing

The earliest source of Thoroughbred sprinting speed was the Irish Hobby (Michael F. Cox, M.D., *History of the Irish Horses,* Dublin, 1897, pp. 22–58). Records of racing in Ireland go back at least 3,000 years. The first Irish ruler dated with any certainty was the Milesian King Heremon whose reign began in 1015 B.C. (P. W. Joyce, *A Social History of Ireland,* Dublin 1920). Three epic poems (in Gaelic) about the Dedannians, the colonists preceding the Milesians, record horse racing, namely the *Voyage of Bran,* the *Romance of Maildune,* and the *Second Battle of Moytura* (P. W. Joyce, *Old Celtic Romances,* 3rd ed., 1907, p. 122). In the latter:

> certain visitors arriving at a meeting were asked if they had hounds [coin] and horses [eich] for races, for when a body of men went to another assembly it was the custom to challenge them to a friendly contest—the hounds to a coursing match and the horses to a race.

Also recorded are races at Tara fair in the sixth century B.C. and at the Curragh of Kildare, still Ireland's principal racing center, during the reign of King Conari I in the first century A.D.

In 441 A.D., at the request of St. Patrick, the laws of the pagan Brehons (judges) were compiled in the *Senchus Mor,* a codex which contains a number of references to racing. (*Brehon Laws, Vol. II,* pp. 153, 155, 161,

Vol. III, pp. 255, 263) As there were then no institutions of learning other than monasteries, the sons of chieftains and of prosperous farmers, in order to further their education, were frequently entrusted to foster parents, for a fosterage fee, payable in cows. The Brehon Laws required that they be instructed in archery, swimming, chess-play, the use of sword and spear, and in horsemanship if they were the sons of chieftains (not farmers). The foster father had to supply horses, including horses for racing at the fairs.

In the *Book of Leinster*, the poet Fulartach (c. 1000–1047 A.D.) notes that the people of Ossory, a sub-kingdom, had a special day at the Fair of Carman in southern Kildare for the "steed contest of the Ossorians." Another horse fair was at Ballinasloe, which is as famous today as ever.

Early Irish races were held at the fairs because of the great crowds which they attracted. Since they were crowd pleasers, these races were run over short courses visible to all spectators. Irish Hobbies were, therefore, sprinters, bred for short speed for more than a thousand years. It was their bloodlines which made the first contribution of short speed to the Thoroughbred.

Description of Irish Hobby Speed (1517–1618)

The word 'hobby' corresponds to the English Dobbin, the Scottish Hoby, the French Haubin, the Italian Obino and the Latin Hobelarius. Its use dates well back. In 1296 A.D. the Irish cavalry serving in Scotland were called 'Hobelarii' because they rode Hobbies.

Hobbies were bred and raced by rival nobles and chieftains in Ireland. An illuminated manuscript in the British Library illustrates the invading English army, 1399, led by the Earl of Gloucestershire, confronting Art Macmurchada, King of Leinster, riding a white Irish Hobby of great quality "which had cost him, it is said, four hundred cows, so beautiful and good it was" (Harleian Manuscript, 1399).

In 1565 Thomas Blundeville said:

> The Irish Hobby is a prettie fine Horse, having a good head & a bodie indifferently well proportioned, saving that many of them be slender & pin buttocked [prominent hip bones], they be tender mouthed, nimble, light, pleasant, and apt to be taught, and for the most part they be amblers, and therefore very meet for the saddle, & to travell by the way: yea, and the Irish men both with

darts & with light spears, do use to skirmish with them in the field. And many of them do prove to that use verie wel, by means they be so light and swift." (Blundeville, Thomas, *The foure chiefest Offices belonging to Horsemanship*, London, from the 1609 reprint of the 1565 edition)

He also recommended Hobbies for such extreme exercises as "to gallop the buck or to follow a long winged hawk."

The Earl of Gloucestershire (left) at the head of the British army during King Richard II's campaign in County Wicklow, 1399 A.D. Approaching at the gallop on the right is Art Macmurchada, King of Leinster, on a white Irish Hobby, "a horse without saddle or seat which had cost him, it was said, four hundred cows, so beautiful and good it was." (From the Harleian Manuscript, *1399)*

In addition to Blundeville, there were many others who testified as to the speed of the Irish Hobby. In 1517 the Bishop of Armagh in a report on Ireland to Henry VIII said: "The land itself produces absolutely nothing but oats and most excellent, victorious horses, more swift than the English horses." The Latin poet Paulus Jovius in his *Historica Descriptio Hiberniae* (1548) wrote:

> Hardly does the horseman need boots; the delicate stepping of the horse keeps him from the mud. There is moreover something magnificent, a kind of majesty in his whole frame, which exalts his rider with pride as he outstrips the wind in his course.
> (John Major, *The History of Greater Britain*, translated from the 1522 Latin editions by A. Constable, Edinburgh, 1892, pp. 53–54.)

Richard Stanihurst in his *Description of Ireland* (1577) distinguishes:

1) the Irish Hobbie "of pace [amble] easie, in running wonderful swift,"
2) Horses of [military] Service "they amble not, but gallop and run," [and]
3) "You shall have of the third sort, a bastard or mongrell [i.e. cross-bred] hobbie-strong in travelling, easie in ambling and verie swift in running."

John Dymmok said (1599):

> The cuntry yeeldeth great store of beefes and porkes, excellent horses of a fine feature and wonderful swyftnes, and are thought to be a kinde of the race of Spanish [pacing] Genetts. (*A Treatice of Ireland* [British Museum, Harleian Ms. 1291], Dublin, Irish Archaeological Society, 1842)

In *The Perfection of Horsemanship* (London, 1609), Nicholas Morgan wrote:

> The Irish Hobby [is] a Horse of middle size, comely and well shaped, and of much courage and fury.... These Hobbies are... naturally inclined to ambling, and carrie their heads so proudly, (having most commonly the benefits of high crests, and deepe necks) and their bodies so comely, that fewe Horses tread with greater state and majestie; they doe naturally delight to clime hilles, and will runne against the same with great fierceness and swiftness.

In addition to the Thoroughbred, today's descendants of the Irish Hobby include the Connemara pony, the American Quarter Horse and several of the American gaited strains.

The revolt of Ireland's nobles and chieftains collapsed in 1583 when their young leader, the Thirteenth Earl of Desmond, was killed. Most of their lands and livestock were given by Queen Elizabeth I to English settlers. The stud farms, which for over a thousand years had bred Hobbies with such success, ceased to exist. The remaining stock drifted about in semi-feral herds. Gervase Markham (1568–1637), who disliked all "forraine" horses, described these wanderers in his 1617 book *Cavalarice, Or the English Horseman* (Book I, pp 16–17):

> Next and last, I place the Irish Hobbie which is a horse of reasonable good shape. Having a fine head, a stronge necke and a well cast body; they have quicke eyes, good limbs and tollerable buttocks: of all horses they are the surest of foot and nimblest in daungerous passages, they are of lively courage and very tough in travell, onely they are much subject to affrightes and boggards [spooks]. For the most part they are bredde in wilde parts and have neither communitie of the fellowship with any man till they come to the saddle which many time is not till they come to seaven, eight, nine or ten years olde, at what time the country rysing [gathering] dow forcible drive the whole studd, both horses, mares, colts and fyllies into some bogge, where being laide set, they halter such as they please, and let the rest goe.
>
> This wylde bringing up, and this rude manner of handling, doth in my conceite engender this fearfulness in the beast, which those ruder people know not how to amend. This horse, though he trots very well, yet he naturally desireth to amble.

Irish Hobby Exports

Exports of Hobbies to Italy for Palio Racing (1461–1538)

During the fifteenth and sixteenth centuries, Irish and English Hobbies were much sought after to compete in the palio races, some of them without riders, between gorgeously caparisoned horses. Staged with great pomp and display, the races, representing the various quarters (Contrada) of Italian cities—Modena, Ferrara, Mantua, Sienna, Florence, Bologna, Verona, Asti, Rome—were run in the main streets or city squares.

Speed and the Thoroughbred

A palio race in the streets of Florence, Italy, 1418, by Giovanni di Francesco Toscani (c. 1370–1430), part of the decoration of a cassone, tempera and gold on canvas.

Correspondence between the Italian nobles, their ambassadors and emissaries, and the British sovereigns and other officials appears [in translation] in the published *Calendar of State Papers, Milan, 1385–1618*. In July 1461 Count Ludovico Dallugo, sent to England by Francesco Sforza (1401–1460) Duke of Milan to buy Hobbies, returned via Calais with "nine hackneys of the kind called Obi [Obini, Ubini, Hobbies], all dappled, but not too big" (*Calendar of State Papers, Milan, 1385–1618*, p. 100). On Jan. 5, 1471, Galeazzo Maria Sforza, Duke of Milan (1444–1476), instructed his Aulic Councillor, Francisco Salvatico, about to go to England:

> We desire you to obtain some fine English hackneys of those called obi for the use of ourself and the duchess our consort, as well as some greyhounds for our hunting—we are sending you el Rosseto, our master of the horse, and two of our dog keepers who know our tastes and the quality of horses and dogs that we require.

Unfortunately Salvatico was unable to carry out his mission because, on the way to England, he was captured and imprisoned by the Duke of Burgundy.

The Archives of Modena record special envoys of the Dukes of Ferrara sent to England in order to secure Royal permission to go to Ireland to buy Hobbies [Ubini] for their masters. Ercole d'Este (1433–1505), Duke of Ferrara, was an enthusiastic breeder of Irish Hobbies, who sent his emissary, Biasio de Birago, to three British sovereigns asking for permission to go to Ireland to buy them—in 1470 to Henry VI, in 1479 to Edward IV, and in 1498 to Henry VII. Biasio

Alfonso D'Este (1470–1534), Duke of Ferrara by Titian.
The Duke imported Hobbies from Ireland in 1527 and 1528.

succeeded in bringing back Hobbies (equos obinos) from all three missions, including twelve in the autumn of 1498 (Cox, pp 23–26).

Alfonso d'Este (1476–1534) shared his father's interest in Hobbies. On November 3, 1527, he sent to Ferrara's ambassador to England, Hieronimo Fiefofino, a request for two Irish Hobby stallions and eight mares "of easy pace, soft and agreeable."

Ludovico, the Duke's falconer, and Master Hannibal, his Farrier, were sent from Ferrara to Ireland (*Calendar of State Papers, Venice and*

Northern Italy, Vol. VI, Part III, 1884, pp.1608–15). They returned empty-handed because "the population there is... in arms, one against another." Returning to England, they were able to buy for twenty-three crowns a bay Hobby stallion "lately arrived from Ireland and of handsome presence, his paces soft and agreeable." Perhaps so as to acquire stock suitable to produce horses for palio racing, they went to the racehorse county of Yorkshire where they located eight Hobby mares for sale "gold dappled, good goers, of handsome shape," three of them Irish. King Henry VIII made the Duke of Ferrara a present of these mares, and added a dark brown Irish Hobby stallion "which is a good goer." They left England with the horses on September 15th, 1528 (Cox, pp. 26–33). Further comments on this purchase appear in a later chapter.

Francesco Gonzaga (1466–1519), Marquess of Mantua, who married Isabella d'Este, daughter of Ercole, received the following.

> November 15, 1511. POLYDORE VERGIL TO THE MARQUIS OF MANTUA. In pursuance of his letters he has bought eight hobbies and sends them by his servant Simon. They go through France with the French ambassador's passport. Good horses are scarce in England, where they are spoilt by being worked too young.

These Hobbies in fact were a gift from Henry VIII.

In 1514, Giovanni Ratto was sent to London with a present of four horses to Henry VIII. In a letter to his wife (October 1514) Francesco noted the arrival of the British Ambassador with a return present of "bellissimi cavalli Ubini." On Jan. 8, 1532, Henry VIII wrote the Marquess Federigo Gonzaga (1500–1540) that he was sending him "duos ex nostratibus gradarios equos." The court of Mantua in the sixteenth century was renowned throughout Europe for art, architecture, literature and music. It included Raphael, Leonardo, Titian, Guilio, Romano, Ariosti, Castiglione and Monteverdi. The Gonzagas continued to breed Hobbies; a classified inventory of the stud in 1690 lists eleven Ubini. (Carlo Cavriani, "Le Razze Gonzaghesche di cavalli nel Mantovano e la loro influenza sul puro sangue inglese" in *Rassegna Contemporanea*, Marzo, April 3, 1909.)

Hobbies of Henry VIII

Henry VIII was a keen sportsman who maintained packs of buck hounds, hart hounds, otter hounds and beagles, and a stable of racehorses. His Privy Purse expenses during the November 1529 through December 1532

King Henry VIII of England (1509–1547) on an ambling palfrey. Contemporary woodcut.

(*Privy Purse Expenses of King Henry VIII*, edited by Nicholas Harris Nicolas, London, Pickering, 1827, pp. 18, 28, 39, 119, 132, 162, 224) contain ample details of his sporting pursuits. His racing stable of "rynning gueldings" at Greenwich Palace was "surveyed" by Thomas Ogle, "Gentleman Rider of the Stables," with the help of attendant jockeys called "rynning boyes." The accounts record scores of payments to Ogle for these boys, particularly for board, boots, shoes, hose (quartered and particolored), coats, doublets and night caps. Fees for riding races are also recorded: On January 16, 1530, 40 shillings "paide in reward to Ogull and the ij boyes for rynning the kings gueldings"; and on April 16, 1530, 20 shillings "to Mr. Halle's servant that kept the white nag", and 10 shillings "to the boye that Ranne the same nag." Many of the great nobles shared with the Irish people the passion for horse racing which continues to this day. The King procured Irish Hobbies for his racing stable. As set forth in earlier paragraphs, the King's enthusiasm for Irish Hobbies preceded these Privy Purse accounts. He made presents of Hobbies to the Marquesses of Mantua in 1511, 1514 and 1532, and to the Duke of Ferrara in 1528. Entries in the Privy Purse accounts note that servants of "Maister Skevington" were paid 40 shillings "for bringing hawks out of Irlande" and £3 "for bringing iij hobbyes to the kings grace" on March 5, 1530, and September 14, 1531. The servant of the Bishop of Armagh (John Kite) "that brought ij hobbies to the king" on January 14, 1531, received 40 shillings. The king's Master of the Horse, Sir Nicholas Carew, bought Hobbies from a horseman with the typically Irish name of George Henyngham (Hennigan). On March 27, 1531, he paid him the sum of 46 pounds "for one hobby and ij gueldings" and on May 6, 1531, board at the rate of 2 1/2 pence a week for "a dounce [dun] hobby." An entry in the June 1532 Privy Purse accounts reads: "Item, the xx daye, paied to a servant of my lorde of kyldare in rewarde for presenting of a couple of hobyes to the King at grenewiche [Greenwich] 40 shillings."

The donor ("my lorde of kyldare") was Gerald FitzGerald, Ninth Earl of Kildare, Viceroy of Ireland. Sir Theodore Cook notes that the Irish peers Gerald FitzGerald, Ninth Earl of Kildare (died 1539), and Barnaby FitzPatrick, Ninth Earl of Ormand and Ossory (died 1581), kept large studs for the breeding of Hobbies which they raced also in England (Sir Theodore A.Cook, *History of the English Turf,* London, 1901, Vol. I, pp. 26, 32). In his *History of Newmarket* (1886, Book II, p. 62), Hore records a fragment of the manuscript stud book of Lord Kildare in the British Museum, which cannot now be located.

Henry VIII also imported Hobbies from his estates in Ireland, noted by Peter Edwards in *The Horse Trade of Tudor and Stuart England* (1991, p. 28).

Other Exports to the European Continent

In 1534, immediately after the false report of the death of his father, "Silken Thomas" FitzGerald, Tenth Earl of Kildare, feared he would be dispossessed by Henry VIII. He sought to enlist the help of Charles V,

The Viceroy of Ireland, Gerald FitzGerald, Ninth Earl of Kildare, school of Holbein.
He presented two Irish Hobbies to King Henry VIII in 1532, which were added to the King's racing stable at Greenwich Palace.

the Holy Roman Emperor and Archduke of Austria, by sending him a present of "fourteen fair Hobbies."

In the *Calendar of State Papers, Henry VIII*, it is recorded that in 1546 the King presented to the Queen of France, "Hobbies, greyhounds, hounds and great dogs," which made her "the gladdest woman in the world."

Exports from Ireland to Virginia (1666)

The incessant warfare between rival chieftains and their British governors led to the extinction of the Irish Hobby. A revolt of the nobles ended in 1583, when its young leader, the Fifteenth Earl of Desmond, was killed.

Speed and the Thoroughbred

Engraving (1850) of Castle Mattress (now known as Castle Matrix), Rathkeale, County Limerick, Ireland, from which Sir Thomas Southwell in 1666 sent a Hobby stallion and four mares to his friend, Sir William Berkeley, Royal Governor of Virginia, at Jamestown.

His lands, which included most of southwest Ireland, were parceled out by Queen Elizabeth I to deserving military leaders and to court favourites. The Southwell family of Castle Matrix in County Limerick was among the few recipients who continued to breed from the Hobbies they found on the acquired estates. In 1666 Sir Thomas Southwell of County Limerick sent from Castle Matrix, via Cork, a Hobby stallion and four Hobby mares to the Green Spring Plantation, near Jamestown, of his friend, Sir William Berkeley, Royal Governor of the colony of Virginia.

The Irish Hobby

Sir William Berkeley, Royal Governor of Virginia 1641–1652, 1660–1677.
In 1666 he purchased an Irish Hobby stallion and four mares from his friend, Sir Thomas Southwell of Castle Mattress, County Limerick, Ireland, and shipped them from Cork in the spring of that year to his Green Spring Plantation near Jamestown, Virginia. Their sprinting bloodlines were one of the principal sources of speed in the foundation of the American Quarter Horse.

Governor Berkeley never recognized the British Commonwealth (1642–1660), Oliver Cromwell nor Parliamentary rule. Many cavaliers emigrated to the "Old Dominion" where their sprinters provided speed for sport and comfortable gaits for travel and plantation over-seeing. Match races on parallel paths at a quarter of a mile, documented as early as 1672, became the most popular form of public entertainment. Governor Berkeley's Irish Hobbies and other sprinter shipments provided foundation bloodlines for the present-day American Quarter Horse. The earliest sources of sprinting speed in the Thoroughbred and in the American Quarter Horse are identical.

Speed and the Thoroughbred

Green Spring, painted by the eminent architect, Benjamin Latrobe (1764–1820), watercolor (1797).
The mansion was built in the mid-seventeenth century by Governor Sir William Berkeley on his plantation 2 miles from Jamestown, Virginia.

CHAPTER III

HOBBY INFLUENCES IN ENGLAND

In Sir Theodore Cook's classic *History of the English Turf* (1901) the earliest ridden horse races cited are from the reign of King Richard I (1189–1199 A.D.), namely races at London's Smithfield Market described by William FitzStephen (d. c. 1190).

The Helmsley Stud (c. 1548–1687)

A major stud with a Hobby background, whose stallions and mares are registered in the *General Stud Book*, was the Helmsley stud of the Rutland family, not far from Richmond, one of the great estates of the North Riding of Yorkshire. The Rutlands' love of racing is documented from as early as July 15, 1549. The Second Earl's Grey Markham lost a race at Berwick that day. The late Major John Fairfax-Blakeborough, author of four volumes on northern British turf history, in Vol. III (p. 16), comments:

"The Earls of Rutland were amongst the most prominent breeders of bloodstock at the very outset of the experimental evolution [of the Thoroughbred], and we know that they raced hobbies at York in the sixteenth century, and used mares of this breed as a foundation for producing bigger and speedier racehorses."

Fairfax-Blakeborough cites the following receipt:

> Receaved, the xxth day of March, 1596, fortie-five pounds wone upon your Lordship's hobbie, of Mr. Holmes [at] the Forest race [course] of Galteresse the xxj of Februarye.

This receipt was directed by his Receiver General, Roger Bayne, to the Fifth Earl of Rutland (1576–1612) who succeeded to the title February 21, 1587. In 1590 when he was 14 years old, with the approval of Queen Elizabeth's chief Minister, Lord Burghley, the young Earl left Corpus Christi College, Cambridge, and went back to Belvoir Castle for the duration of the hunting season. A thorough knowledge of hunting was part of the essential education of a young nobleman with a great estate. The Belvoir hounds today, still the property of the Duke of Rutland, are among the world's most famous packs of foxhounds.

The Fifth Earl loved horse sports and he loved the theatre. During the 1590s, he frequently joined his friend, the Third Earl of Southampton, at the playhouses and at the taverns frequented by actors and authors. Shakespeare, who dedicated *Venus and Adonis* to Southampton (1593), was well known to Rutland.

The Fifth Earl continued to maintain his stud, with its Hobby bloodlines, until his death in 1612. In 1605, he sent three horses to race at Royston. He was succeeded (June 26, 1612) by his younger brother, Francis Manners, the Sixth Earl (1578–1632), also a racing man. The Rutland manuscripts record his horses racing at Lincoln and at Newmarket in 1617, 1618 and 1621.

The Dowry of Lady Catherine Manners (1620)

On May 16, 1620, the Sixth Earl's daughter, Lady Catherine Manners, married George Villiers, First Duke of Buckingham (1592–1628), who had been appointed Master of the Horse in 1616 by James I. Lady Catherine's dowry included 1,000 pounds sterling in cash, the Helmsley estate and its livestock in North Yorkshire. This included the stud of brood mares, many with Hobby bloodlines—a magnificent gift.

Buckingham was an expert horseman with a flair for dressage, acquired during his early education in France (1610–1614). His patronage of horse racing was limited. On March 8, 1622, at Newmarket, in a race for £100, his horse Prince was beaten by Lord Salisbury's horse. Two days later, when he backed Sir John Shelley's horse with £100, the winner was Lord Fielding's horse (J. B. Muir, *Old Newmarket Calendar*, 1892, p. 17).

Buckingham's influence on the evolution of the British racehorse has been much exaggerated. During the 8 years of his marriage (he was assassinated August 23, 1628), he seems to have taken little part in the management of the Helmsley stud. The Duke's own bloodstock was kept at his Highe Wair stud. An inventory drawn up in 1623 listed twenty-two

Francis Manners, Sixth Earl of Rutland (1578–1632), succeeded to the title in 1612. In 1620 his only child and heiress, Lady Catherine Manners, married the First Duke of Buckingham, Master of the Horse, bringing as part of her dowry the Helmsley estate in North Yorkshire and its stud of racing stock.

The right Honorabell FRAVNCIS MANNERS Earle of Rutland Baron Rofs of Ham: lake Beluoire and Trufbutt and Knight of the Honorable order of the Garter.

mares, of which five were Spanish, three were Barbary, and fourteen were apparently bred in England (J. B. Muir, *Frampton and the Dragon*, 1895, pp. 106–108). A list of thirty-nine mares at the Royal stud of Tutbury, made in 1624 when Buckingham had been Master of the Horse for 8 years, showed only six were covered by "The Arabian Colt"; the balance were covered by dressage sires, namely Neapolitan Coursers, Spanish Jennets and "The French Horse." As a horseman, Buckingham's primary interest was not in racing, but in his fine figure on horseback, his

Speed and the Thoroughbred

George Villiers (1592–1628), First Duke of Buckingham, attributed to Gerrit van Honthorst (1590–1656).
George Villiers, appointed Master of the Horse in 1616 by James I, was created First Duke of Buckingham in 1623. This portrait, attributed to van Honthorst of Utrecht who came to England in 1625, shows Buckingham, an accomplished dressage rider, in a slashed white silk doublet and hose on a colorful Barb stallion with red trappings.

handsome face, his gorgeous clothes, his equally gorgeous bridles and saddles with their gold brocade housings, the subtlety of his aids as well as in the beauty and spirit of the horses he rode, and the ease and proficiency with which they performed "Airs above the Ground," such as seen today at the Spanish Riding School of Vienna.

Catherine Manners, the Duchess of Buckingham, did more for the British racehorse than her famous husband. During the 8 years of marriage (1620–1628) when they were constantly at court, the Helmsley stud with its Hobby bloodlines probably was carried on in the same fashion as during the years when it was the property of her uncle and of her father, the Fifth and Sixth Earls of Rutland.

The Duchess loved Helmsley and its racing bloodstock. This explains why it was the major part of her dowry. Her keen interest in and her extensive knowledge of horses was demonstrated after the Duke's death when she took over the distribution of his seventy-seven household horses (Muir, *Frampton*, 1895, p. 110). "Coach horses, Hunting horses, Greate [dressage] horses, pacing nags, breeding mares and colts" were presented to more than twenty-five individuals, from the King to "the Duchess' Dutch picture drawer." This was Gerrit van Honthorst (arrived 1625 in England) who painted a magnificent portrait of the Duke of Buckingham on a gaily colored Barb stallion.

The Second Duke of Buckingham (January 30, 1628–1687) was only a few months old when his father was assassinated (August 23, 1628). During his long minority, his mother was in charge of his affairs. Subject to her approval, the matings of mares would have differed little from long-standing family patterns performed by a capable and trustworthy staff.

For 24 years this great lady and highly competent racehorse breeder, Catherine Manners, Duchess of Buckingham, as owner, wife and guardian, directed the progress of the then most important Helmsley stud. In a society where racing was the exclusive domain of men, she was an extraordinary woman, certainly the first to succeed among the men of the British racing world.

In 1644 Helmsley, one of the most strongly fortified medieval castles in Britain, was besieged, captured and razed by the great Parliamentary general, Thomas, Third Lord Fairfax. The entire estate, including the horses, was sequestered by the Commonwealth. The Second Duke and his mother took refuge in France. During the next 7 years Parliament made no apparent effort to bring about changes at Helmsley or to add to its livestock. Any matings would have been carried out, as customary, by

Speed and the Thoroughbred

**The First Duke of Buckingham (1625)
by Peter Paul Rubens (1577–1640), oil on wood.**
George Villiers, later the first Duke of Buckingham, married Catherine Manners, daughter of the Sixth Earl of Rutland, in 1620, acquiring thereby the Helmsley stud and its racing stock.

The Buckingham family (1628) by Gerrit van Honthorst.
Sitting in the garden are the Duke and Duchess (the former Lady Catherine Manners), their older daughter and the infant son born January 30, 1628. He became the Second Duke when his father was assassinated on August 23, 1628. About 1635, acting as guardian of her seven year old son, the Duchess bred at Helmsley the earliest mare registered in the General Stud Book. *This was Old Bald Peg, dam of the Old Morocco Mare (G.S.B., p. 14).*

the same staff. In 1651, however, Helmsley and its livestock was given to Lord Fairfax by Parliament "as a salve for an old wound" received during the 1644 siege.

From 1651 to September, 1657 the Helmsley stud was owned and operated by Lord Fairfax. It was the Helmsley horses of this period that first appeared in the *General Stud Book*. As one of the great cavalry leaders of history, the primary interest of Lord Fairfax was military horses, not racehorses. His manuscript treatise on horses, published by Lady

Wentworth (1938), deals primarily with cavalry horses. However, although he lived at Nun Appleton, Fairfax had sufficient interest in horse breeding to utilize the racing stock he found at Helmsley.

Helmsley's Old Morocco Barb and Old Bald Peg

Among the horses bred at Helmsley by Lord Fairfax during the period 1651–1657 was Old Peg, also known as the Old Morocco Mare, registered on page 14 of the *General Stud Book*. Her sire was probably the Old Morocco Barb, the earliest imported stallion registered in the *General Stud Book*. Her dam was Old Bald Peg, bred by the Duchess at Helmsley while acting as guardian of her 7 year old son, the Second Duke.

A letter to Lord Deputy Wentworth, dated November 9, 1637, written from London by the Archbishop of Canterbury reads: 'The greatest news is of the Morocco Ambassador, whose present to His Majesty [Charles I] of [four] Barbary horses and saddles of great value, attracted much attention" (Prior, *Royal Studs*, 1935, p. 74; Hore, *History of Newmarket*, 1886, Vol. II, p. 191). These horses were sent to the Royal stud of Charles I at Tutbury in Staffordshire. Their descendants were the five mares named Morocco who are listed in the July 24th, 1649, Tutbury inventory.

Thirty-six brood mares are listed in the Tutbury inventory. Only two stallions would have been required to service a brood mare band of this size. Two were surplus. One of these surplus stallions was probably Helmsley's Old Morocco Barb, sire of the Old Morocco Mare. This mare is registered in the *General Stud Book* as Old Peg, her dam Old Bald Peg, by an Arabian, her grand-dam by a Barb. This pedigree is taken from John Cheny's 1743 *Racing Calendar*. Cheny was not known for accuracy. The 'Arabian' and the 'Barb' were probably imaginary. That Old Bald Peg had some Barb blood was suggested by her bald (white) face, a Barb characteristic.

Of the 78 Thoroughbred foundation mares registered in the *General Stud Book*, Old Bald Peg was the earliest, the most important and the most influential. She appears in the pedigrees of all three "Great Progenitors," Matchem 1748, King Herod 1758 and Eclipse 1764, and consequently in the pedigree of almost every present-day Thoroughbred, often several times. The major sources of sprinting speed in the bloodlines of Old Bald Peg were the Hobby strains, bred by the Earls of Rutland at the Helmsley and Belvoir Castle studs during the sixteenth and early seventeenth centuries.

Thomas, Lord Fairfax, Third Baron of Cameron (1612–1671) engraved by William Faithorne (c. 1616–1691) after a portrait by Robert Walker (1607–c.1660).

Commander (1642–1650) of the Parliamentary armies of Oliver Cromwell, Fairfax captured Helmsley in 1644. As a reward for his services, he was given the entire Helmsley estate and all its bloodstock in 1651. Between that date and 1657, Fairfax bred Old Peg, also known as the Old Morocco Mare.

Descendants of the Old Morocco Mare (Old Peg)

Spanker

Old Bald Peg's daughter, the Old Morocco Mare (Old Peg), was bred at Helmsley when it was the property of Lord Fairfax. She could not have been foaled later than September 1657 when the Second Duke of Buckingham, then 29 years old, secretly stole back from exile and regained his Helmsley Estate by marrying Mary, daughter and heiress of Lord Fairfax. By 1660, having recovered both his riches and his vanity, Buckingham purchased for an enormous price a spectacular suit of clothes, designed to attract more attention than the King, to wear at a reception honoring the Restoration of Charles II to the throne.

When Buckingham bred the Old Morocco Mare to the D'Arcy Yellow Turk (deduced to be by Place's White Turk), she produced Spanker whom the *General Stud Book* (p. 14) calls "the best horse at Newmarket in Charles II's reign" (1660–1685), the best because he was the leading winner of 4-mile multiple heat King's Plates. He was also the leading British-bred sire of the seventeenth century.

Spanker was the first Thoroughbred. Of all the seventeenth century stallions, he alone has a recorded pedigree containing all three speed strains—Hobby, Running-Horse and Turcoman. He stood at Helmsley for several seasons, along with the Helmsley Turk, probably until Buckingham's death in 1687. The *General Stud Book* states that Spanker "was afterwards Mr. [Charles] Pelham's, of Brocklesby, and covered there, and was sometimes described as Mr. Pelham's Bay Arabian." (*G.S.B.*, p. 14)

The Arabian stallion fashion prompted several owners of British-bred stallions to change their horses' names to "Arabian," provided they had some Arabian ancestry (*G.S.B.*, p. 388). Pelham's Arabian (Spanker) was by the D'Arcy Yellow Turk (*G.S.B.*, p. 389), bred by James D'Arcy the Elder. Also by the Yellow Turk was the Oglethorpe Arabian (*G.S.B.*, p. 389), the property of Sutton Oglethorpe, appointed Master of the Studs by Charles II, after D'Arcy's death in 1673. The two horses were called Arabians because their sire, the Yellow Turk, was by Place's White Turk (*G.S.B.*, p. 388). The Yellow Turk was out of a D'Arcy Royal Mare.

Buckingham had the pleasure of cheering his horse to victory in one race after another. He was also the breeder of the important sire, the Helmsley Turk (*G.S.B.*, p. 389), presumed also to be by Place's White Turk.

Hobby Influences in England

The Second Duke of Buckingham (1628–1687) in court dress. An engraving by Robert White (1645–1703) after a portrait by Sir Peter Lely (1618–1680).

In the autumn of 1657, Buckingham secretly stole back to Nun Appleton, the residence of Lord Fairfax. He married Fairfax's only child, Mary, thereby recovering his inheritance, Helmsley and its livestock. When he bred the old Morocco Mare (Old Peg) to D'Arcy's Yellow Turk, she produced Spanker, referred to by the General Stud Book *(p.14) as "the best [race] horse at Newmarket."*

The Effigies of the Most Noble, George Duke, Marquess & Earle of Buckingham, Earle of Coventry, Viscount Vil- ers, Baron of Whaddon & Knight of ye most No- ble order of the Garter

R. White sculp:

Following Charles II's death in 1685, his younger brother, King James II, so detested Buckingham that the Duke retreated to Helmsley. With all his faults, the Duke had great charm, was "condescending" (friendly) to everyone, a fine horseman and a sportsman. Adored by his sheep-farming tenants, he assembled a pack of foxhounds to protect their lambs from the strong Cleveland Hills foxes. These foxhounds are reputed to have been the first major pack which hunted only foxes. In 1687, after a great spring hunt, he contracted a chill so severe that even the best efforts of two girls at each side in bed could not keep him warm. He died shortly after the hunt on April sixteenth.

The North Milford Stud

In addition to Spanker, the Old Morocco Mare produced three fillies: Young Bald Peg, by the Leedes Arabian; a full sister to Spanker; and a filly by her own son, Spanker, known as the Spanker Mare (*G.S.B.*, p. 17). The *General Stud Book* assigns Young Bald Peg to "R. Milbanke" and the other two fillies to "Lord D'Arcy." This last (and incestuous) breeding suggests the Old Morocco Mare may have remained at Helmsley. Young Bald Peg, by the Leedes Arabian, was acquired by Edward Leedes (born c. 1638) of the North Milford stud. Edward (Englebert) Leedes inherited North Milford in the parish of Kirby Wharfe, south-east of York, near Tadcaster, on the death of his father Robert (1596–1656). He was a major market breeder, selling beautifully bred fillies and other bloodstock descended from his two foundation mares to many prominent racing men in different parts of the country.

In the seventeenth century, the Thoroughbred was a small regional North Yorkshire breed. Of the 78 earliest mares (mostly late seventeenth century) listed on pages 1 to 19 of the 1891 edition of the *General Stud Book*, 73 were on stud farms within a 30-mile radius of the North Yorkshire town of Richmond. Several new or expanded stud farms were located near Newmarket, to which Charles II had moved the center of racing after his Restoration in 1660. Leedes' sales fostered the evolution of the Thoroughbred from a regional to a national breed.

After Edward Leedes' death January 29, 1703, the stud was continued by his son Anthony (died June 1711) and then by Anthony's younger brother, Robert, who carried on until he died in 1767.

Edward Leedes bred Young Bald Peg to her own sire, the Leedes Arabian. There probably were two Leedes Arabians, so-called because they were probably descended from the D'Arcy Yellow Turk (*G.S.B.*, p. 389). The elder Leedes Arabian, bred to the Old Morocco Mare and to

North Milford, home of the Leedes family, in the parish of Kirby Wharfe, southeast of York, near Tadcaster. When Edward Leedes bred Young Bald Peg (Leedes Arabian ex Old Peg [Old Morocco Mare]) to her own sire, she produced Bay Peg. Sold to Sir William Ramsden of Byram Hall, Bay Peg bred to the Byerley Turk produced Basto, foaled c. 1702.

Young Bald Peg, flourished in the late seventeenth century, the property of Edward Leedes. Anthony Leedes probably stood a younger Leedes Arabian, whose portrait was painted by John Wootton (1683–1764).

To the cover of the elder Leedes Arabian, her own sire, Young Bald Peg produced Bay Peg (*G.S.B.*, p.37), a bay filly, whom Edward Leedes sold to Sir William Ramsden of Byram Hall, near Ferrybridge, Yorkshire. In 1701 or 1702, Ramsden bred Bay Peg to the Byerley Turk, a tail male descendant of Place's White Turk. The result of this mating was Basto, foaled 1702 or 1703 (Pick, *Turf Register*, 1803, Errata page).

From 1708 to 1710, Basto won several match races, including five at Newmarket which are unpublished. He was then sold as a stallion to the Second Duke of Devonshire and stood at the Duke's estate, Chatsworth, in Derbyshire, where he died in 1723.

The Leedes Arabian by John Wootton (1683–1764), signed and inscribed "The Arabian horse belonging to Mr. Leedes."

There were probably two horses known as the Leedes Arabian. He first appears in the General Stud Book as the sire of Bay Peg out of Young Bald Peg, also by the Leedes Arabian out of the Old Morocco Mare (Old Peg) who was foaled at Helmsley between 1651 and 1657. The younger, the probable subject of the Wootton painting, was bred by Edward's son, Anthony Leedes, between 1703 and 1711.

Bay Peg's next recorded foal was born 12 years after Basto. This was Fox, a bay colt by Clumsy, born in 1714, and bred by Sir Ralph Ashton, Bart., who apparently had purchased Bay Peg from "Mr. Leedes" (Anthony or Robert). One of the early eighteenth century's best racehorses, Fox won eleven races, including a match for 2000 guineas, and was a most successful stallion.

Brocklesby Betty

Brocklesby Betty, a chestnut mare, foaled in 1711, the property of Charles Pelham of Brocklesby Park, Lincolnshire, was a brilliant racehorse (1716–1718). James Weatherby, compiler of the *General Stud Book* (1791) said she was "thought to be superior to any horse or mare of her time." By Curwen's Bay Barb, Betty was bred by Henry Curwen of Workington Hall, Cumberland, who had acquired her dam, the Hobby Mare, from her breeder "Mr. Leedes" (Pick, *Turf Register,* 1803, p. 5). The dam of the Hobby Mare was Edward Leedes' Piping Peg. The *General Stud Book* (p. 11) mistakenly identifies Piping Peg as a grey mare, the property of the Duke of Kingston. The Kingston mare was racing in 1706 (Prior, *Royal Studs,* 1935, p. 193); she could not have been the grand-dam of Brocklesby Betty, foaled 1711. No pedigree for Leedes Piping Peg is provided, either by the *General Stud Book* or by Pick. However, the bloodlines of the Leedes' brood mares included Old Bald Peg, Old Morocco Mare (Old Peg), Young Bald Peg and Bay Peg, the last two owned by Mr. Leedes. It is evident that he named his mare Piping Peg because her pedigree also traced back to the Helmsley stud's Old Bald Peg.

The third foal of the Old Morocco Mare (Old Peg), a full sister to Spanker, had only one recorded foal, Lord Lonsdale's Counsellor by the Shaftesbury Turk.

The Spanker Mare

The Old Morocco Mare's fourth foal, registered as the Spanker Mare (*G.S.B.*, p. 17), was more prolific. Incestuously bred by Lord D'Arcy to her own son Spanker, the Old Morocco Mare produced the Spanker Mare who was added to the brood mare band of the North Milford stud by Edward Leedes.

The Spanker Mare's five foals were all by the elder Leedes Arabian. Her second foal was the successful sire, Leedes, whose portrait was also painted by John Wootton. Her third foal, registered as the Leedes

Arabian Mare (*G.S.B.*, p. 12), was known as Cream Cheeks—the garish head markings, characteristic of Barb bloodlines, undoubtedly derived from the double cross in her pedigree of the Old Morocco Barb.

Fortunately, Edward or Anthony Leedes bred Cream Cheeks to Leedes Old Careless, before selling the mare to Sir William Strickland, who bred her to less distinguished stallions. The product of her first mating was Betty Leedes. The latter's sire, Old Careless (*G.S.B.*, p. 378), by Spanker out of a "Barb" mare, was bred in 1692 by Edward Leedes and sold to the Fifth Lord Wharton (1648–1715), of the Winchendon stud, the host of the Quainton race meeting in Buckinghamshire.

Old Careless was the best Newmarket racehorse of 1698. In 1699 or 1700, Lord Wharton returned Careless to Edward Leedes in Yorkshire where the stallion would be bred to top-class mares, including Cream Cheeks. The pedigree of the latter's first foal, Betty Leedes, contained all three Thoroughbred speed strains, including three close crosses of the great matriarch Old Bald Peg. Betty Leedes was acquired by Leonard Childers of Carr House, near Doncaster.

Flying Childers

William, Second Duke of Devonshire, a keen racing man, maintained one of England's most important studs of bloodstock at Chatsworth, his magnificent Derbyshire residence. Succeeding to the title in 1707, he purchased, in 1710, the noted racehorse and sire, Basto, to head his band of brood mares.

In 1719, although he had several young racehorses of his own breeding, the Duke recognized the potential of an untried 4 year old bay colt whose pedigree was rich in Helmsley bloodlines, and promptly purchased him. From his later portraits we know that in appearance this was an average sort of horse with a plain head, a high set tail, and good legs and feet.

Five years earlier, Leonard Childers had taken Betty Leedes from Carr House, Doncaster, to Aldby Park, seat of the Darley family, to be covered by the Darley Arabian, imported in 1704 from Aleppo by way of Smyrna. It was the bloodlines of this mare, Betty Leedes, which induced the Duke of Devonshire to purchase from Leonard Childers her 4 year old colt. He was known thereafter as Devonshire or Flying Childers.

The Duke of Devonshire's gamble paid off handsomely! The *General Stud Book* (p. 379) said Flying Childers was "the fleetest horse ever trained in this or any other country." His racing record is set forth in full

Flying Childers, attributed to James Seymour.
Flying Childers (Darley Arabian ex Betty Leedes) was bred by Leonard Childers of Carr House, near Doncaster. Flying Childers was purchased by the Duke of Devonshire, for whom he won every time he started.

by C. M. Prior in his *Royal Studs* (p. 120). After his racing days were over (c. 1723) Flying Childers went to stud at Chatsworth, primarily as a private stallion. He died there in 1741.

Bartlett's (Bleeding) Childers

Flying Childers' younger full brother, Bartlett's (Bleeding) Childers, never raced due to his infirmity. He was sold to Mr. Bartlett, who stood him for public service at Nettle Court, near Masham, Yorkshire. He covered many good mares, and was the great-grandfather in tail male of Eclipse 1764. This tail male line of the Darley Arabian through Bartlett's (Bleeding) Childers is today the male line of perhaps 90% of all Thoroughbreds throughout the world.

Why did the Duke of Devonshire value so highly the bloodlines of Betty Leedes? The pedigree of Betty Leedes contains three close crosses to Old Bald Peg, a foundation mare of the Helmsley stud whose stock traced back to the Hobbies bred there by the Earls of Rutland in the sixteenth century. It was this intense concentration of sprinting bloodlines which gave Flying Childers his phenomenal speed.

During the seventeenth and eighteenth centuries, most British breeders placed great emphasis on the sire, and far less on the dam. The Duke of Devonshire was more astute. He appreciated fully the importance of female bloodlines in general and the superiority of the Hobby sprinting strains bred by the Earls of Rutland. Born in 1673, he could perhaps, as a child, have seen Spanker run at Newmarket. He was certainly familiar with Spanker's outstanding record as a sire. In 1710 he had bought Bay Peg's son, Basto, as a stallion to stand at Chatsworth, where the horse was then located (1719). He was impressed with the races won (1716–1718) by Piping Peg's granddaughter, Brocklesby Betty. In 1719 at York, Basto's half-brother Fox (Clumsy ex Bay Peg) won the Ladies Plate for 5-year-olds—4 miles, carrying 10 stone (140 lbs)—beating eleven horses.

The final proof was Flying Childers. Old Bald Peg's eminent male descendants in tail female include: Spanker, her grandson c. 1675, "the best [race] horse at Newmarket during Charles II's reign" (1660–1685) and leading British-born sire of the seventeenth century; the sensational racehorse and important sire Flying Childers 1715; and the latter's full brother, Bartlett's (Bleeding) Childers, tail male great-grandsire of Eclipse 1764.

The Black Barb Mare

Another foundation tap root mare was the Duke of Rutland's mare by Mr. Massey's Black Barb (*General Stud Book*, p. 9). J. B. Robertson (*Origin and History of the British Thoroughbred Horse*, 1940, p. 32) says: "She went back to one of the Belvoir Castle [Helmsley stud] Running Horse strains."

Although the *General Stud Book* does not record her pedigree, this was evidently well known to the Second Duke of Devonshire, who acquired her two daughters, both by his stallion Basto. In a trial over the "New Round Course" at Newmarket, he matched the older filly, Brown Betty 1713, against Flying Childers who left her far behind (Pond, *Sporting Calendar*, 1751). The young filly, Old Ebony 1714, presented him with five foals by Flying Childers. The Duke of Devonshire bought Basto and Flying Childers because they were descended in tail female from Old Bald Peg. Undoubtedly for the same reason, he acquired the daughters of the Rutland mare by Mr. Massey's Black Barb.

Tail female descendants of the Black Barb Mare include: Hermit 1842; the French invader, Gladiateur 1862, who won both the Epsom Derby and the Doncaster St. Leger; Doncaster 1870; Triple Crown winner Ard Patrick 1899; and the great American sire and Kentucky Derby winner Native Dancer 1950.

CHAPTER IV

Running-Horses

The Racehorses of Jervaulx

One of the most shameful episodes in English history took place during the years 1535–1540 when Henry VIII seized and destroyed the abbeys and monasteries of the Roman Catholic church. In 1535, backed by the *Act of Supremacy* enacted the previous year, the King appointed Thomas Cromwell (1485?–1540) Vicar General to carry out the desecration. In the same year a survey and evaluation of all ecclesiastical property was prepared, the *Valor Ecclesiasticus*, a remarkable achievement. In 1536 Cromwell carried out this destruction of the smaller monasteries, the value of whose properties was less than £200. In the autumn of 1536 and through 1540 the destruction of the larger monasteries was completed. Cromwell, falling from favor, was executed July 28, 1540.

Of special interest to racehorse breeders was the Abbey of Jervaulx in North Yorkshire, situated on the River Eure, in Wensleydale, near the town of Ripon, site of a famous horse fair founded in the Middle Ages. Its extensive lands, its magnificent church and many monastic buildings were assessed at £455, in the *Valor Ecclesiasticus*. Founded by Irish monks in 1156 A.D., Jervaulx was one of several abbeys of the Cistercian order whose large landholdings were devoted to livestock, primarily to sheep, and to grain growing. In these ventures they had the unpaid help of many "lay brothers" who occupied separate buildings. These were adjacent to the church, library, refectory, kitchens and monks' quarters.

Stables and sheds completed the compound. This was surrounded by a massive wall, more than a mile long. The number and size of the buildings, outlined by surviving foundation walls, indicate their occupation by several hundred monks and lay brothers.

A letter dated June 8, 1537, addressed to Thomas Cromwell, the Vicar-General, was written by Sir Arthur D'Arcy, great-grandfather of James D'Arcy the Elder (1617–1673) of the Sedbury stud. The text is as follows:

> It shall like your Honourable Lordship [Thomas Cromwell] to be advertised that I was with my Lord Lieutenant at the suppression of Gervaix [Gervayes, Jervaulx, when the monks were driven from the monastery]; which house within the gate is covered [roofed] wholly with leads; and here is one of the finest churches that I have seen, fair meadows and the river running by it, and a great domain [tract of land].
>
> The King's Highness is at great charge with his studs of mares at Thornbery and other places which are tyne [fen, marshy] grounds, and I think that at Gervaix and the Granges incident [farm houses with farm buildings in other parts of the property] with the help of their great hardy commons [community grazing fields] the Kings Highness, by good overseers, should have there the most best race [stud] that should be in England, hard and sound of kind.
>
> For surely the breed of Gervaix for horses was the tryed [racing] breed in the North, the stallions and mares well sorted. I think in no realm should be found the like to them, for there is hardy and good high grounds for the summer, and in winter woods and low grounds to fire [warm] them ... from Gervaix I went to Sallay. (*The Ruined Abbeys of Yorkshire*, W. C. Lefroy, 1891, p. 178; *The Royal Studs*, C. M. Prior, 1935, p. 3)

This letter has many facets. Shortly before it was written (June 8, 1537), Sir Arthur D'Arcy had witnessed the heart-rending spectacle of monks and lay brothers being forcibly ousted from their beloved monastery, occupied by their Cistercian order for over five centuries. Later in June, Adam Sedburgh, 18th Abbot, was taken to the Tower of London and hanged. D'Arcy's opening description of the beauty and majesty of the monastic buildings, of the granges, the pastures and the river, indicates his horror that the glory of Jervaulx was about to be destroyed. His letter to Cromwell, suggesting the King occupy it as a stud farm, was a desperate appeal that it be spared. This was not to be. A letter to Cromwell headed "At York this 14th day of November, 1538," signed by Richard Bellycys, in charge of the demolition, records the melting down of the lead roof tiles, bids for the church bells, and plans to raze the walls.

The high quality of Jervaulx's pastures, described by D'Arcy, confirms the eminence of North Yorkshire as the center of racehorse breeding during the sixteenth and seventeenth centuries. Most importantly, the D'Arcy letter provides what is, perhaps, a unique glimpse of selective racehorse breeding during the fifteenth century. It reads: "For surely the breed of Gervaix for horses was the tryed breed in the north [of England], the stallions and mares well sorted." In the sixteenth century, the word race meant stud, and the word tryall meant race. In D'Arcy's letter, the words *'tryed breed'* meant 'best racing breed.' The words *'stallions and mares well sorted'* means that selected stallions were bred to selected mares, rather than turned out with the mares and mated indiscriminately.

When D'Arcy described *'the breed of Gervaix for horses,'* he used the past tense. Why and when did the monks start breeding racehorses? The Abbey was founded by Irish monks. The Cistercians kept excellent records. Had Cromwell not ordered the demolition of Jervaulx, we would be able to provide accurate and extensive answers to these questions. Raising racehorses was undoubtedly profitable. The monks had pasture, grain and lay brothers to do the work. Top-class racehorses were bred at Jervaulx during the fifteenth century and possibly earlier. Since the Abbey was in the horse breeding center, the market for these horses was close at hand.

What happened to the Jervaulx bloodstock? The horses were gone on June 8, 1537 when D'Arcy wrote. They would not have left before 1535 when Cromwell took over. Perhaps the answer will be found in other D'Arcy correspondence, yet to be discovered.

What became of the Jervaulx *"well sorted"* racing stallions and mares? Did Cromwell try to sell them, along with the church bells and the lead from the roofs? Were they sold to a single owner, such as an Earl of Rutland, owner of the nearby Helmsley stud farm? Were the horses sold in groups or individually? It seems unlikely that their bloodlines were lost. Jervaulx was surrounded by the stud farms of other racehorse breeders who would have made every effort to secure these famous and precious bloodlines.

Sir Arthur D'Arcy's letter of June 8, 1537, gives us a tantalizing glimpse of selective racehorse breeding and racing in the fifteenth century. To turf historians, Thomas Cromwell met a fitting fate when Henry VIII "hanged him by the neck."

Town Plates and Bells

During the sixteenth and seventeenth centuries short distance sprint racing flourished and expanded in England and Scotland. Many towns offered annual races, open to all, in which the prize was a silver bell or 'plate' (covered cup, tankard, etc.). For example: Richmond 1512, Gatherly Moor course (Fairfax-Blakeborough, *Northern Turf History*, Vol. I, 1948, p. 171), York 1530, Forest of Galtres course (ibid., Vol. III, 1950, p. 15), Chester 1540, Haddington 1552, Kiplingcotes 1555, Richmond 1570 (town course), Croyden 1574, Ayr 1576, Salisbury 1585, Doncaster 1595, Carlisle 1599, Huntington 1602, Paisley 1608, Brackley 1612, Durham 1613, Stamford 1620, Tarporley 1622, Little Budworth 1622, Lanark 1626, Harleston 1632 and Hyde Park, London, 1636 or earlier. The town of Croyden also provided a trophy in the form of a golden snaffle bit, donated by the Earl of Essex (d. 1600).

Later town-organized race meetings could also be cited. These races were organized by the city fathers. Besides entertaining the townspeople, they attracted crowds from the surrounding countryside who went to church and patronized shops, merchants, inns and taverns. These courses were also used in match races of 4 to 12 miles between noblemen and gentlemen riding their "Hunting-Horses." Because of the length of these races, spectators saw little of the excitement. Most of the annual races organized by the cities and towns were run over short distances, such as a quarter of a mile, in full view of everybody, from start to finish.

The popularity of these crowd-pleasing short races for "Running-Horse" sprinters with speed bloodlines is illustrated by William Camden's account of the City of York race (*Britannia*, 1586).

> The Forest of Galteresse [course] is famouse for a yearly horse-race where the prize is a little golden bell. It is hardly credible how great a resort of people there is to these races, from all parts, and what great wagers are laid.

'Forest' was the name of a Royal park-like game preserve, not a wooded tract. This course was the center of British racing until the 1660s when Charles II moved it to Newmarket after his Restoration. During the Middle Ages, the Forest of Galtres was a royal game preserve where red and roe deer, wild boar, wolves and bear were hunted. It extended from York about 20 miles to the town of Aldburgh.

By founding King's Plate races (1665) and by his frequent appearances on the heath, Charles II made Newmarket the center of British racing.

The previous center was North Yorkshire, the Forest of Galtres course (before 1530), near the city of York, and the Gatherley Moor course (c. 1512) near the town of Richmond, in the center of the studs that were breeding racehorses. The pastures of this region were among the best in England.

Gervase Markham (1568?–1637)

Running-Horses and Running-Horse racing were fully described by Gervase Markham of the Manor of Cotham in Nottinghamshire. He was the first Englishman to write a book about horses who was an accomplished horseman. The book was entitled *A Discource of Horsemanshippe*, published in London by Richard Smith and entered in the Stationer's Register (John Charlwood) January 20, 1593 (E. N. L. Poynter, *A Bibliography of Gervase Markham*, Oxford Bibliographical Society 1962, pp. 216, illus.). Markham's predecessors had been translators and compilers. The Renaissance of Arts and Letters during the fifteenth and sixteenth centuries was centered in Italy. These included the art of horsemanship. In 1550 Federico Grisone published in Naples *Gli Ordini di Cavalcare*. This was the first book to describe the patterns of riding now known as dressage, seen today in competition from beginners to the Olympic Games, and at the Spanish Riding School of Vienna.

The Earl of Leicester, Queen Elizabeth I's Master of the Horse, encouraged Thomas Blundeville to translate Grisone's book into English; the translation was published in 1560. The Earl also brought to England in 1566 the Italian, Claudio Corte, whose 1562 book, *Il Cavalerizzo* was translated by Bedingfield and published in abridged form in 1584. Blundeville's massive work, 1565, was titled *The Four Chiefest Offyces Belonging to an Horseman—Painfully Collected From a Number of Authors*.

Markham was familiar with these books and their instructions which he cites in his texts. The purpose of the *Discource* was quite different. Markham's book is a manual for training racehorses. The subtitle of its first edition reads in part:

> Also the manner to chuse, trayne, ride and dyet both Hunting-Horses [stayers] and Running-Horses [sprinters] with all the secrets thereto belonging discovered. An article heretofore never written by any author.

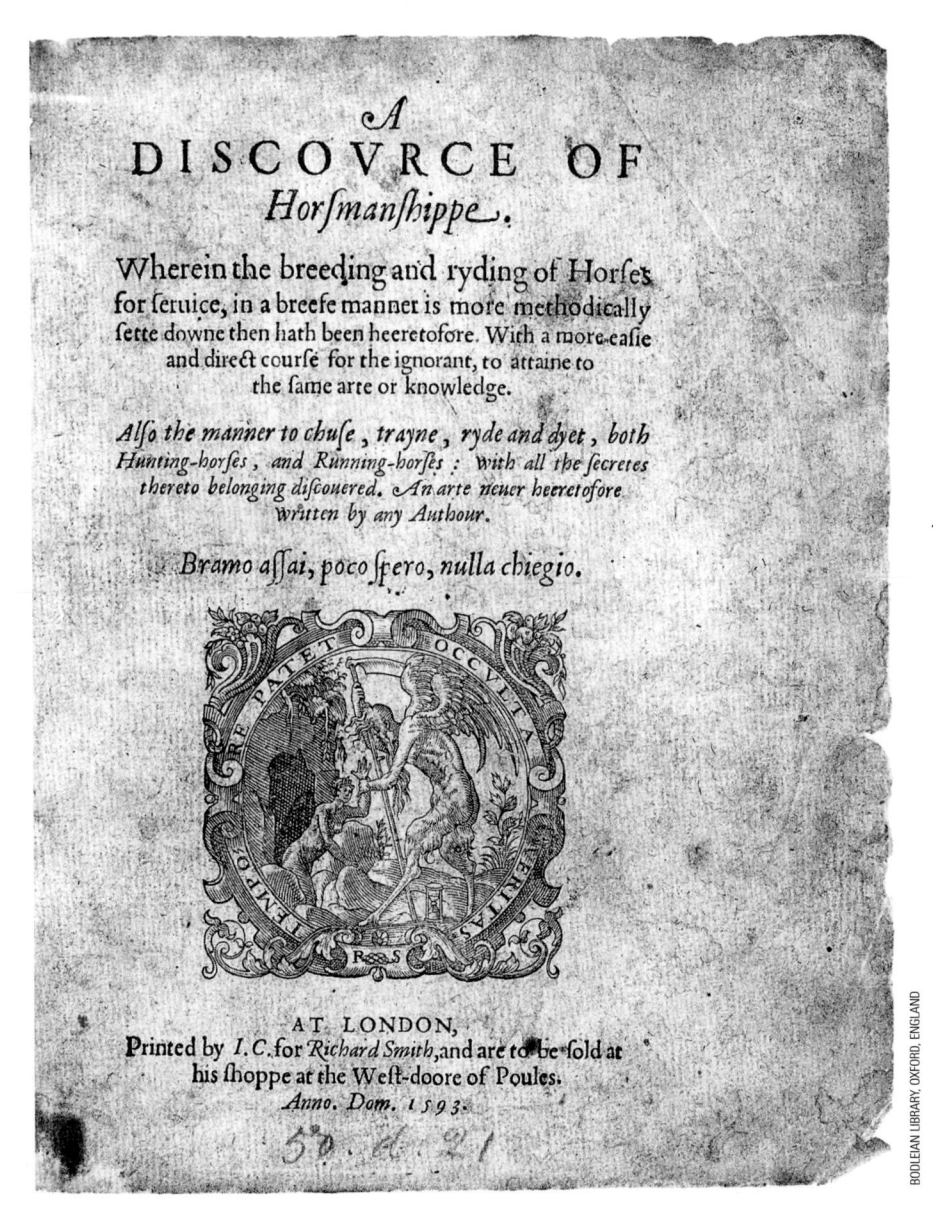

The first (1593) edition of Gervase Markham's treatise on training racehorses—the manner to chuse, trayne, ryde and dyet, both Hunting-horses, and Running-horses—*probably the most popular text on training racehorses ever printed. From 1593 to 1734, with this and other titles and under Markham's name, it was published 29 times in London. This was also the first American horse book, published in Wilmington, Delaware in 1764 and in other editions as late as 1842.*

Markham's claim was fully justified. The 1593 *Discource of Horsemanshippe* was the first book on training racehorses ever published. To emphasize its principal purpose, the titles of the second and later editions were changed to the wording of the subtitle as quoted above.

Markham's manual on training racehorses, his first book, was a bestseller, with six editions dated 1593, 1595, 1596, 1597, 1599 and 1606. The complete text on Running-Horses, under the heading "the Sixt Booke," was included in the 1607 and 1617 editions of Markham's larger work, *Cavalarice, or The English Horseman*. Throughout the seventeenth

Title page of the fourth edition (1609) of The four chiefest Offices belonging to Horsemanship *(Breeding, Riding, Shoeing, Veterinary Science) compiled by Thomas Blundeville from European works. The first edition was published in 1565.*

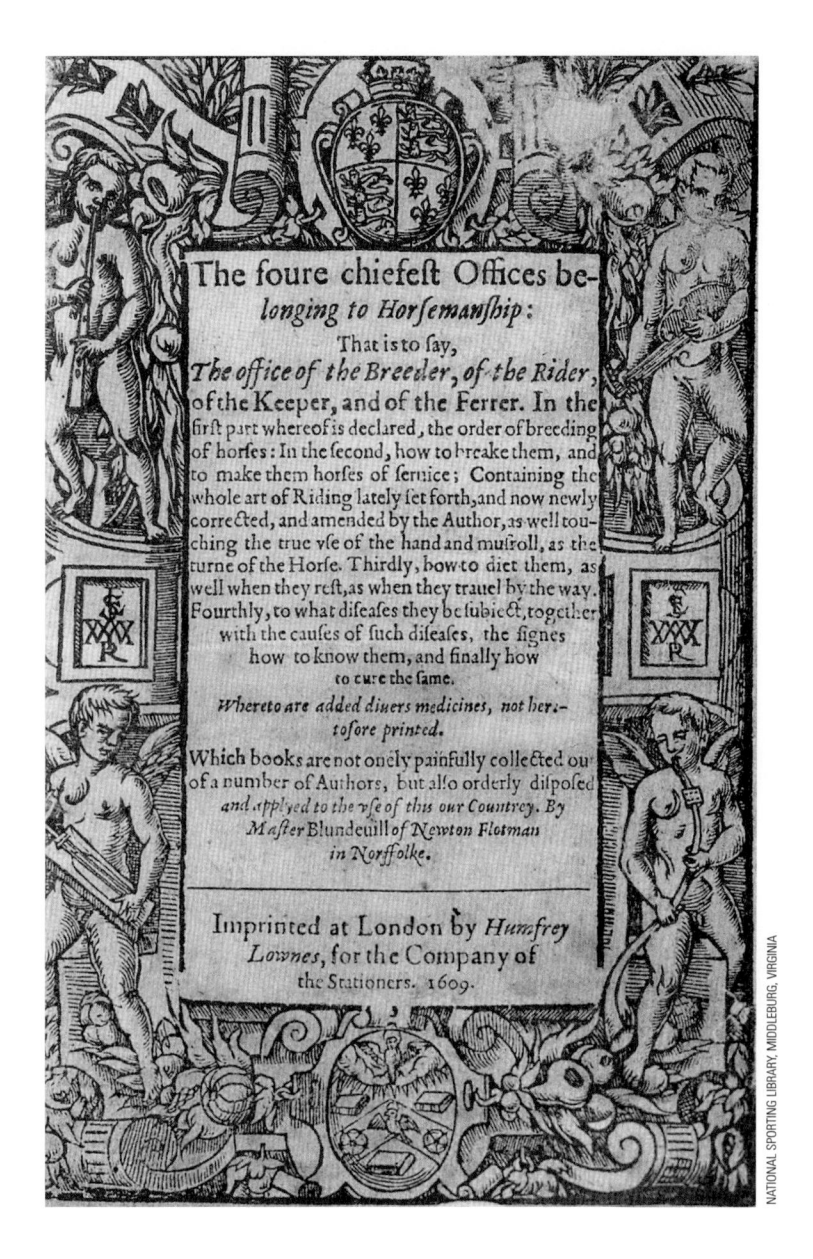

century the multiplicity of annual town race meetings, primarily for speedy Running-Horses, created a widespread demand for the book.

This demand was probably a surprise to the author. In 1593, he was only 25 years old, uncertain about his future career. In 1593 he was attached to the household of Shakespeare's first patron, the Third Earl of Southampton, boon companion of the Fifth Earl of Rutland. The principal interests of these two young noblemen were horse racing and the theatre. Shakespeare and Markham would have been well-acquainted.

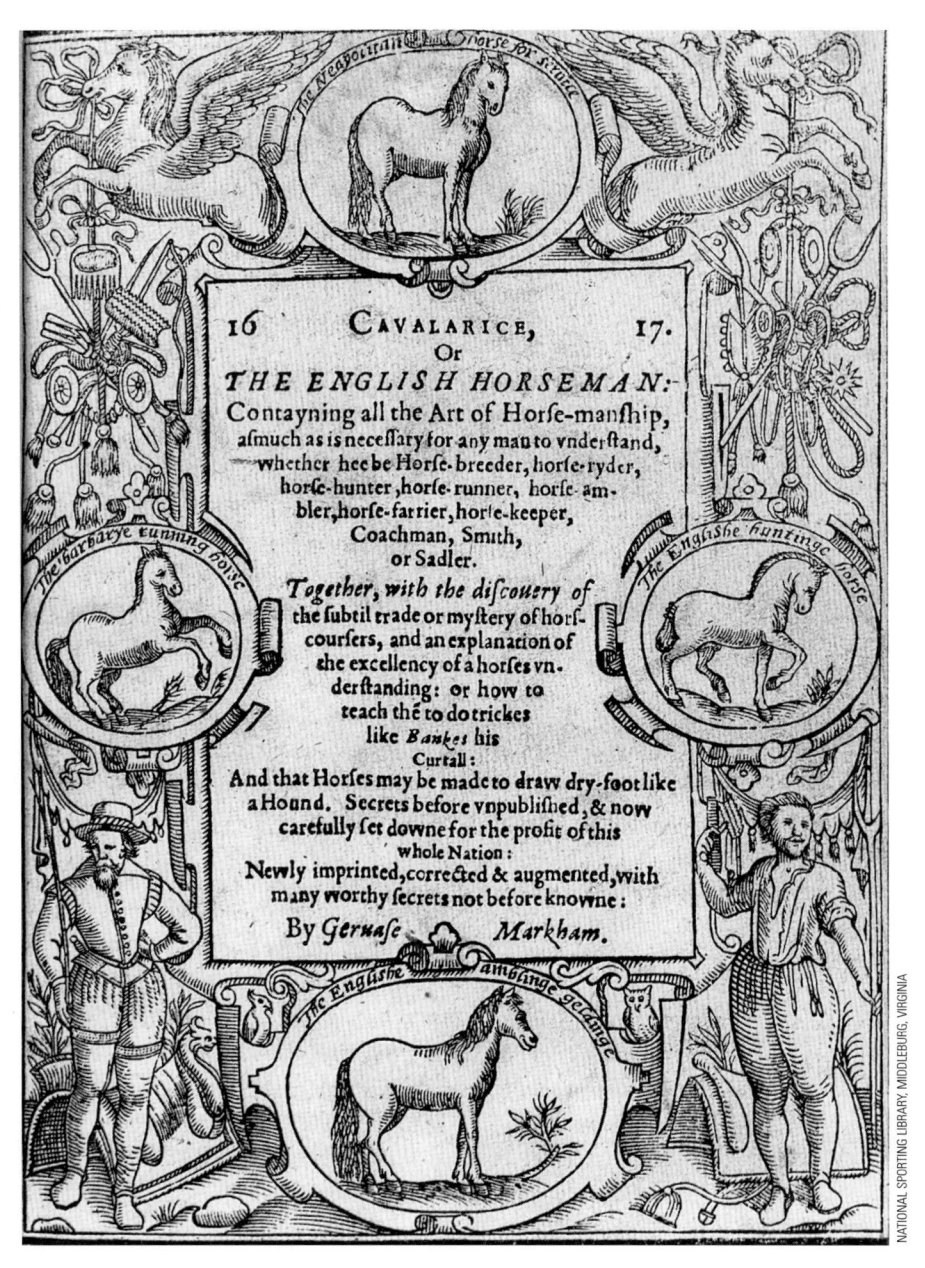

The main title page of the 1617 edition (first edition, 1607) of Cavalarice *by Gervase Markham, an accomplished horseman. The border depicts four types of horses used in England and shows a riding master with his wand and a groom standing beneath trophies of spurs, harness, hay forks and curry combs.*

Gervase Markham was a brilliant rider who truly loved horses, who treated them with "uncompromising kindness," and who was born with a natural understanding of their thoughts, emotions and physical limitations. He was "not a scholar to men and fashion, but onlie to experience and reason." His first instructor was his father, Robert Markham (1536–1606) also a noted horseman, to whom the 1593 *Discource* was dedicated. Concerning his second instructor, he speaks of a certain man:

> how he had laboured with his horse for two years with such an instrument of torture [the bit] to no avail, when the same Horse, being brought to the most famous Gentleman, and worthy all praise-full memorie, Maister Thomas Story of Greenwich [the horse] was by him in lesse than one halfe yeare made the most principall, best doing horse which came upon the Black-heath, and my selfe at that instant riding with him, did so diligently observe both his art, his reasons, and his practise, that even from that man and that Horse, I drew the foundation and ground of my after [later] practise. (*Cavalarice*, 1607, Book II, p.25)

Markham trained a "blacke bastard courser" for the personal use of the Earl of Essex, the Queen's favourite courtier. This was a half-bred horse, by a Neapolitan Courser stallion, the Italian breed of dressage horse. In 1589 he trained a "pied [piebald] Markham," sold to Sir Henry Sidney and later acquired by the French Ambassador. In 1600 Markham entertained the Queen and the court with an exhibition of horsemanship (dressage).

Shortly before his death (April 6, 1589), Sir Francis Walsingham, the Secretary of State, through diplomatic channels, was able to import a superb stallion "from a parte of Arabia called Angelica, to Constantinople, and from thence to the highermost partes of Germaine by lande, and so by sea to England." The horse was presented to Robert Markham. Being unbroken, his education was entrusted to Robert's son, Gervase. In his 1593 book, Markham speaks with enthusiasm of the horse "of which I have the ryding"—and is "even now under mine hands." His detailed description of the horse is a classic.

How to Train Racehorses: Background of the Discource

Markham's 1593 *Discource of Horsemanshippe* was the definitive text on training racehorses throughout the seventeenth century and well into the eighteenth. How and where did this 25 year old young rider accumulate the wealth of practical information and the detailed yet simple instructions which made the book so popular? Throughout the Middle Ages and into the sixteenth century, England's great nobles lived chiefly on their estates, in houses so vast that they required a host of servants. Many of these were young men of good birth who, in addition to their tasks, acquired the social polish and the accomplishments required by their station in life.

Cotham, the residence of the Markhams, was only 4 miles from the immense Belvoir Castle, the seat of the Earls of Rutland, where Gervase became a *Gentleman Serving-Man*. His country background and horsemanship undoubtedly led to Markham's early employment in a variety of outdoor sports. This is confirmed by Markham's *A Health to the Gentlemanly Profession of Serving Man,* published May 15, 1598. The occupations stressed are "hunting, hauking, fishying and fowlying."

Markham described himself as having "a covetousness for knowledge." He was a diligent and keen observer and listener. During his term of service with the Rutlands he had many opportunities to meet, to talk and to listen to owners, trainers and jockeys, to breeders, stud managers, grooms and farriers. Gervase Markham made the most of these opportunities. The information so gathered was transferred to the pages of his 1593 *Discource of Horsemanshippe*

The Helmsley Stud

The Rutlands kept their hunting horses at Belvoir Castle. The Belvoir foxhounds continue to this day, the property of the Dukes of Rutland. Their racing stock was kept at the Helmsley Castle stud, not far from Richmond. As noted in the previous chapter, Irish and English Hobbies were the most coveted of the several strains of speedy sprinters grouped under the heading *Running-Horses*.

Hunting-Horses and Running-Horses

Markham describes two types of racing popular during the sixteenth and early seventeenth centuries. Stag hunting (red deer) and buck hunting (roebuck) were sports reserved for noblemen and gentlemen. With their owners in the saddle, horses ridden to hounds ran in match races of 4 to 12 miles, backed by wagers for large sums of money. The horses bred for this type of racing were known as *Hunting-Horses*. At the opposite end of the scale were short distance races for fast sprinters, open to all, from noblemen to commoners, in which participants and onlookers alike delighted, and on which large sums were also wagered. The horses bred for this type of racing were known as *Running-Horses*. Markham distinguishes the stayer for long distances which he calls the Hunting-Horse, from the sprinter which he calls the Running-Horse. His description reads:

> The Hunting Horse hath his virtue consisting of long and wearie toyle, this other [Running-Horse] in quickness of speede and suddaine furie. And as the one requireth the whole day for his tryall, so this other in comparison must dispatch in a moment. (In Elizabethan times, *tryall* meant race, and *race* meant stud.)

Markham elaborates:

> Now you shall know, that for as much as the Hunting Horse, and the Running Horse are for two serverall ends, that is to say, the first for long and wearie toyle by strength and continuance of labour, struggling and working out his perfection: the other by suddaine violence, and present furie, acting the uttermost that is expected from them: therefore there must necessarilie bee some difference in the ordering and dyeting of these two creatures: the Hunting Horse by strength, making his winde indure a whole dayes labour; the Running horse by winde and nimble footemanship, dooing as much in a moment as his strength and power is able to second.

It has been stated that the bloodlines of the Hunting-Horse and of the Running-Horse are identical, that we are considering one strain of horses, not two. Markham did not agree. He wrote:

> These [Running-Horses] of which I have hitherto spoken, being of great courage and mettall, are intended to be of great speede and swiftnesse, for it is impossible to finde toughness and furie joyn'd to gether, because the one doth ever confound the other. (*Cavalarice*, 1617, Sixt Booke, pp. 3, 46)

The Running-Horse of the sixteenth and early seventeenth centuries is the second source of Thoroughbred sprinting speed, the first source being the Hobby (Chapters II and III).

Sprinting Running-Horses were more numerous and more popular than staying Hunting-Horses. Since only noblemen and gentlemen were allowed to take part in buck and stag hunting, the match races between their Hunting-Horses were small in number. There were many more races open to Running-Horses. Short sprint races took much less energy and effort than 4 to 12 mile multiple heat match races for Hunting-Horses. Thus, a Running-Horse could race more frequently than a Hunting-Horse.

These conclusions are supported by the Duke of Newcastle. In his book *A New Method and Extraordinary Invention to Dress Horses* (London, 1667, p. 80), in the chapter headed *"Few Good Horses,"* he laments the scarcity of:

> winter hunting geldings to gallop or run upon all grounds, with a snaffle [bit], the reins slack upon his neck, which makes him very much the safer for his rider because he gallops on his haunches [hindquarters]. Running-Horses are the most easily found and of the least use [for hunting]; commonly they run upon the heaths (a green carpet) and must there run all-upon the shoulders [forehand], which in troublesome grounds is ready to break one's neck. Though I like the sport [of] a Running-Horse very well. I think myself as Good a *jockey* as any, and have Ridden many Hundreds of *Matches*.

The Duke of Newcastle, in writing "Running-Horses are the most easily found," documents their superiority in numbers and popularity. Of special interest are Newcastle's comments on how racehorses and hunters should gallop. As the old saying goes, "There's nothing new under the sun." Thoroughbred owners and trainers today want their racehorses to gallop on the forehand. Owners, riders and trainers of the seventeenth century also wanted their Running-Horse sprinters to gallop on the shoulders. Present-day foxhunters who ride Thoroughbred horses to hounds object to the low "daisy-cutting" action which results from galloping on the forehand. They prefer more balanced action in which the hindquarters dominate. Newcastle insisted on the same type of action at the gallop across country.

Temperament
Markham emphasizes the importance of temperament in a Running-Horse:

> If your delight may sway you to the exercise of this sport you shall be very carefull in chusing a horse for your purpose where the chiefest thing to be regarded is his spirit—for his spirit would be free and active, inclined to cheerfulness, lightness and forwardness to labour, scope or gallop. For a horse of a dull, idle and heavie nature can never either be swift or nimble. (*Cavalarice*, 1617, Sixt Booke, p. 2)

Conformation of the English Running-Horse
Markham continues:

> The true English Horse—him I mean that is bred under a good clime, on firme ground, and in a pure and temperate zone—is of tall stature and of large proportions; his head, though not so fine as with the Barbaries or Turkes, yet it is leane, long and well fashioned, his crest is hie, only subject to thickness if he be stoned [a stallion], but if he be gelded, then it is thin, firm, and strong; his chyne [back] is straight and broad, and all his limbs large, lean, flat, and excellently joined in them, exceeding any horse of any country whatever. For swiftness what nation hath brought forth that horse which hath exceeded the English?

For a Running-Horse "which hath great speede in short courses" (a quarter of a mile), Markham recommends that:

> …your Running-Horse be somewhat long and loosely made, that is to say somewhat long filletted between the hucklebones [points of the hip] and the short ribbes; if he have slender limbes, long joyntes, a thinne necke, and a little bellie, being in all his generall partes, not so strong and knit together as the Hunting Horse. (*Cavalarice*, 1617, *Sixt Booke*, p. 3)

Quoted earlier in this chapter is the Duke of Newcastle's statement that he liked "the sport [of] a Running-Horse very well … and have Ridden many Hundreds of *Matches*." Newcastle was also an extensive breeder of Running-Horses (Chapter V). His advice on the selection of Running-Horse brood mares *(A New Method and Extraordinary Invention to Dress Horses,* I, London, 1667, 1667, pp. 61–62) reads:

> For English Mares, there are none like them in the World to Breed On: but then you must Chuse them fit for such Horses as you would breed; if you would have mares to breed Running-Horses of, then they must be shaped thus: as light as possible, large and long, but well-shaped, a short back, but long sides, and a little long-legged; their breast as narrow as may be, for so they will gallop, the lighter and nimbler, and run the faster; for the lighter and thinner you breed for galloping is the better.

Seventeenth Century Racehorse Training: Amble, Pace, Rack and Trot

During the sixteenth and seventeenth centuries, it was generally believed that no horse could attain his maximum speed at the gallop unless he was also proficient in the lateral gaits, either in the two beat amble, also known as the pace, or in the four beat rack, also known as the singlefoot. If not naturally gaited, the horse must be taught. The slow amble or rack were used as muscle builders while warming up or cooling out the racehorse. The trot was deplored. Most Running-Horses were natural amblers. If a seventeenth century racehorse was a natural ambler or racker, this indicated a Running-Horse and/or Hobby background.

Seventeenth century books on training racehorses specified amble and rack. These included Markham's *Discource of Horsemanshippe* published in England under various titles 29 times between 1593 and 1734. In 1764, parts of it were reprinted in London and published in Wilmington,

Delaware, then a British colony, the first American book devoted solely to horses. The title page reads: "*By Gervase Markham—and Discreet Indians*"! Similar instructions regarding rack and amble were published in *Browne-His Fiftie Yeares Practice,* London 1624; Michael Baret's *Hipponomie,* London 1618; and Sir William Hope's *Supplement of Horsemanship,* Edinburgh 1696.

"*The Manner to Chuse, Trayne, Ride and Dyet Both Hunting-Horses and Running- Horses*" is the subhead of Markham's 1593 *Discource.* He and the other authors cited insist that racehorses must be fed precisely at regular intervals around the clock. The major item in their diet was especially baked horse bread. In preparation for an important match or race, horses were only worked every 3 or 4 days. For conditioning racehorses, sweats under heavy wool blankets were considered essential. During the past five centuries, trainers have praised and adopted several different systems for making horses run faster. Fortunately the Thoroughbred racehorse, an adaptable animal, has survived them all!

Markham's Racing Reminiscences

In the 1607 and 1617 editions of his book *Cavalarice or the English Horseman* ("Cavalarice" is spelled "Cavelarice" in the 1607 edition), Gervase Markham includes personal reminiscences of short distance sprint racing between Running-Horses, Hobbies, "plain-bred English" horses, Barbs and Turks:

> For swiftness, what nation hath brought foorth that horse which hath exceeded the English? For proofe whereof we have this example: when the best Barbaries that ever were in my remembrance were in their prime, I saw them overrunne by a Black hobbie at Salesburie of Maister [Thomas] Carlton's and yet that hobbie was more overrunne by a horse of Maister Blackstone's called Valentine, which Valentine neither in hunting [distance racing] nor running [sprint racing] was ever equalled, yet was a plaine-bred English horse by both syre and dam. To descend to our instant in time, whatever men may report or imagine, yet I see no shape which can persuade me that Puppie is any other than an English horse: and truly for running [sprinting] I hold him peerless. Again, for infinite labour and long indurance, which is easiest to be seen in our English hunting matches, I have not seen any horse able to compare with the English horse.

Left: An English Running-Horse type. Etching by Wenceslas Hollar (1607–1677) after a drawing in India ink by Francis Barlow (1626–1702).
The engraving was published as Plate 6 of the book, Variae Quadrupedum Species, *1663.*

Speed and the Thoroughbred

Four Running-Horse type sprinters racing at Dorsett Ferry near Windsor Castle, 1687 engraving after a drawing by Francis Barlow (1626–1704).

The sprinters have the characteristic conformation of the early English Running-Horse described by Markham (1593–1617)—efficient for sprinting speed over quarter of a mile racing courses but lacking in quality. The engraving states that this was "the last horse race run before Charles II." Notwithstanding the fact that Charles II had inaugurated middle distance racing (King's Plates) on Newmarket Heath, the illustration suggests that, when he stayed at Windsor Castle where the terrain was unsuitable for 4-mile courses, he continued to enjoy his first love, sprint racing.

The paragraph reveals Markham's dislike of the Barbaries, alleged to carry North African bloodlines, most of them purchased in Egypt, Italy, Marseilles (France) or Spain, Britain's traditional enemy. (Remember the Armada!) This dislike was shared by other contemporary writers: Nicholas Morgan (1609) and Michael Baret (1618). Markham was one of the first to point out that the imports were slower than the native horses. In fact, there is no evidence that any of these Barb, Turk or Arabian stallions, exported from their country of origin, has ever beaten a British horse in sprint races on the race course.

When Markham compared Mr. Blackstone's Valentine, a "plaine-bred English horse, by both syre and dam," with "Maister Carlton's black Hobbie," he pointed out that the latter was from a strain whose speed bloodlines were much admired. When Markham said that Puppie had the shape [conformation] of an English horse "whatsoever men may report or think," he was undoubtedly referring to those who tried to claim Barb ancestry for the fastest Running-Horse (sprinter) of 1607.

The second edition of *Cavalarice* (1617, Sixt Booke, p. 2) includes the following comparisons of speed between the Spanish Jennet, the Turk, the Barb and the British Running-Horse:

> We finde that the Turkes are much swifter horses than the [Spanish] Jenets, and the Barbaries much swifter than the Turkes, and some English horses or geldings are swifter than either Jenet, Turke, or Barbary. Witness Gray Dellaval, the horse upon which the Earl of Northumberland rode—witness Gray Valentine which dyed a horse never conquered; the Hobbie of Maister Carlton, and at this houre, most famous Puppey [Puppie] against whom men may talke, but they cannot conquer.

CHAPTER V

Sprinting Speed Strains Before 1650

A Quarter Race for James I

James Stuart, who became King of Scotland in 1567 and King of England in 1603, was the first ruler who understood the future potential of Newmarket Heath as the center of British racing. He delighted in short distance sprint racing for which Irish Hobbies were especially bred. Although King James kept dressage horses, his first love was quarter-of-a-mile racing.

In 1591, Brian O'Rourke sent King James "six pair Hobbies and four great dogs" (Irish Wolfhounds). William, son of Lord Leitrim, sent him a dark 'dounce' [dun] Hobby in 1606, shipped from the port of Galway and lodged in the Royal racing stables at Greenwich Palace. James' Queen, the former Princess Anne of Denmark, shared her husband's love of horses. In 1619 she sent six Irish Hobbies to King Louis XIII of France.

When James I visited the city of Lincoln in the spring of 1617, the Mayor and Aldermen, well aware of the King's tastes, organized a race on April 3rd on the heath outside the town. In order to keep back the expected crowds, they put up ropes on either side of the course. The following description of this part of the Royal "progress" is contained in the Note Book of the Lincoln archives:

> On Thursday there was a great horse race on the heath for a cupp where his MA^{tie} [Majesty] was present & stood on a scaffold the citie had caused to be set up, & with all caused the race a quar-

Speed and the Thoroughbred

James Stuart from George Turberville's *Book of Falconry*, Second Edition, London, 1611.
James Stuart, King of Scotland 1567–1625, King of England 1603–1625, loved sprint racing over short courses. In 1591 and 1606, he acquired Hobbies from Ireland, the prime source of Thoroughbred speed. The Tutbury Royal stud became his personal property in 1603. He sent ambling horses to Phillip III of Spain in 1614 and 1623. Sharing the King's love of horses, his Queen, the former Princess Anne of Denmark, sent six Irish Hobbies to Louis XIII of France in 1619.

ter of a mile long to be railed & corded with ropes & stoops on both sides, whereby the people were kept out & the horses w^{ch} [which] ronned were seen faire.

It was politically prudent to be a supporter of sprint racing in the first half of the seventeenth century. During that century, the three major breeders of Running-Horses were Sir John Fenwick (1579–1658) of the Wallington stud in the county of Northumberland, the Earl (later Duke) of Newcastle (1593–1676) of the Welbeck stud in Staffordshire, and King Charles I (1603–1649) of the Tutbury Royal stud, also in Staffordshire.

Facsimile of the Minute Book of the City of Lincoln in England, recording a race at a quarter of a mile run April 3, 1617 to honor the visit of King James I. The entry reads:

"On Thursday there was a great horse race on the heath for a cupp, where his ma^tie was present & stood on a scaffold the citie had caused to be set up, & with all caused the race a quarter of a mile long to be railed & corded with ropes & stoops on both sides, whereby the people were kept out & the horses wch ronned were seen faire."

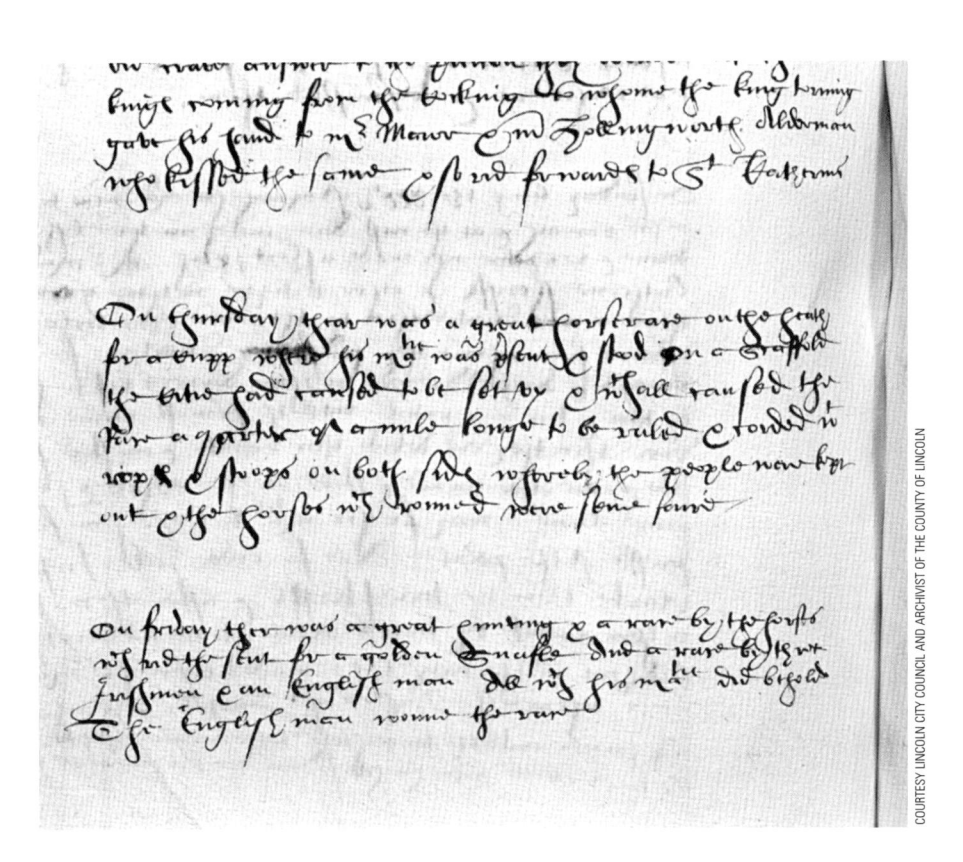

The Wallington Stud of Sir John Fenwick: Running-Horse Strain

In his Chapter X, Section 3, of *A New Method and Extraordinary Invention to Dress Horses* (London, 1667), Newcastle said that Fenwick

> had more experience of Running-Horses than any man in England; for he had more Rare [superior] Running-Horses than all England besides, and the most part of all the Famous Running-Horses in England that ran one against another, were of his Race [stud] and Breed [bloodlines].

Born in 1579, Sir John was 35 years old on September 14, 1614, when the inquest after the death of his father, Sir William Fenwick, was completed. He inherited Wallington and other properties, as well as his father's stud.

Wallington was famous for its lavish hospitality dispensed to horsemen and to their mounts. According to an old North Country proverb, "if you give your horse the bridle, he'll take you to Wallington." The Fenwicks were equally famous for the frequency of their journeys to

Newmarket, and for the speed with which these were accomplished. A nursery rhyme goes: "Fenwick of Bywell's away to Newmarket, and he'll be there before we get started."

Even before he inherited Wallington, Fenwick had become eminent as a breeder of England's fastest racehorses. In 1610 Sir Ralph Delavel wrote as follows to the Earl of Northumberland, whose portrait was painted by Van Dyck:

> I have seen a verye fyne paseinge maire thats blacke and of a myddle syse which I can buy for your lordship, and hath so good a forehand and heed [head], as I knowe not where the like is to be had in these parts. The maire that Sir John Fenwick gave to the King [James I], that was the swiftest horse held to be in Ingland, which was given to the duck of Hulster [Duke of Ulster], that is full sister by the horse to this maire. (Lady Wentworth, *Thoroughbred Racing Stock*, 1938, p. 193)

Charles I made Fenwick a baronet on June 14, 1628. Earlier that year, on March 7, Charles I made William Cavendish the Earl (later Duke) of Newcastle. It was largely through Newcastle's influence that Sir John was appointed "Surveyor of the Race" at the Royal Stud of Tutbury, a post which he continued to hold for 21 years until the execution of Charles I, January 30, 1649.

Sir John Fenwick served seven terms in Parliament, the first representing Cockermouth (1623–1624) during the reign of James I. During the reign of Charles I, representing the county of Northumberland, he served six terms in the "Long Parliament." Faced with the prospect of civil war, Fenwick was so effective in recruiting troops for the King that

Sir John Fenwick (1579–1658) of Wallington, County Northumberland. The Duke of Newcastle in his New Method and Extraordinary Invention to Dress Horses *(1667) said of Fenwick that "the most part of all the Famous Running Horses in England that ran, one against the other, were of his Race [stud] and Breed."*

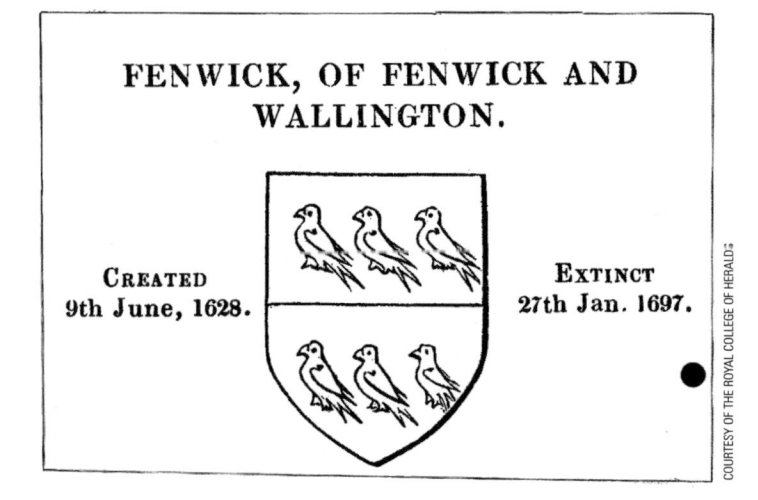

Coat of Arms of Sir John Fenwick, created a Baronet, June 9, 1628, by King Charles I, who also appointed him "Surveyor of the Race [stud]" at Tutbury. Fenwick filled the Tutbury paddocks with Running-Horse and Hobby strain sprinting speed "Royal Mares" which are among the earliest recorded ancestors of the Thoroughbred.

on March 9, 1639, he was appointed Muster-General of the Royal armies. On January 22, 1643, with other members supporting the King, he was excluded from Parliament, but on June 26, 1646, the 67-year-old was reinstated and in the same year appointed High Sheriff of Northumberland.

The Welbeck Stud of the Duke of Newcastle: Running-Horse Strain

Although his world fame rests on his magnificently illustrated treatise on dressage, *Nouvelle Methode et Invention Extraordinaire de Dresser Les Chevaux* (Antwerp, 1657), the Duke of Newcastle was also an expert rider

The Duke of Newcastle on horseback at Bolsover Castle, by Diepenbeke. *An illustration of Newcastle's 1657* Nouvelle Methode.

Speed and the Thoroughbred

Horses of the Duke of Newcastle (c.1593–1676) painted in Antwerp by Abraham van Diepenbeke, a pupil of Rubens.
Engravings after these portraits were published (1657–1658) in Newcastle's Nouvelle Methode et Invention Extraordinaire de Dresser Les Chevaux. *Pictured are (top middle) the gray Nobillisimo, a Neapolitan Courser; (middle row left) Machomilia, light bay, a Turk; (middle row right) Rubecan, black, a Russian (Polish) horse; (front row left) Le Superbe, dark bay, Spanish; (front row right) Paragon, red roan with a bald (white) face, a Barb. The portraits of the Turk and of the Barb are characteristic of the Oriental sires which were bred to native English Hobby race mares during the seventeenth century to increase the distance capacity of their offspring.*

of Running-Horses in short-course races. The text of the first edition of his *A New Method and Extraordinary Invention to Dress Horses* (London 1667, p. 80) declares: "I like the sport of a Running-Horse very well. I think myself as Good a *jockey* as any, and have Ridden many Hundreds of *Matches*."

The Duke of Newcastle was a breeder of Running-Horses. In his section on Spanish horses, he states:

> No Horse is so fit to Breed on, as a Spanish Horse; either for the Mannage, the Warm Ambling for the Pad; Hunting, or for Running-Horses: *Conquerour* was of [by] a Spanish Horse, *Shotten-Herring* was of a Spanish Horse, *Butler* was of a Spanish Horse, and *Peacock* was of a Spanish Mare: And These Beatt all the Horses in their Time, so much, as No Horse ever Run near them. (*A New Method and Extraordinary Invention to Dress Horses,* London, 1667, pp. 50, 51)

The dams of Conquerour, Shotten-Herring and Butler probably were Running-Horse mares from the Duke of Newcastle's Welbeck stud, while the sire of Peacock probably was a Running-Horse.

The Chevalier de St. Antoine

Although his personal favorite equestrian sport was sprint racing, one of the first concerns of the newly crowned King James I was the education of his two sons in Horsemanship (dressage), then an indispensable accomplishment for young men of rank and fortune. He sought help from Henri IV, King of France, who generously offered his Grand Ecuyer Pierre Bourdin, the Chevalier de St. Antoine. St. Antoine had been instructed by Antoine de Pluvinel, whose book *Instruction du Roi*, lavishly illustrated, was the leading work on dressage, published in French and German in 1629.

In 1604, living quarters, stabling and a *manege* (covered arena) were established at the Greenwich Palace mews near the racing stable. The heir to the throne, the brilliant Prince Henry, died of typhoid fever November 6, 1612, but St. Antoine continued as Equerry to Prince Charles, then 12 years old, the future King Charles I. In Anthony Van Dyck's equestrian portrait of Charles, St. Antoine, holding the King's helmet, is shown standing beside the horse.

Many prominent horsemen were instructed by St. Antoine, including Charles I and the Dukes of Buckingham and Newcastle. His *manege* became a popular gathering place for racing men as well. A letter written on November 21, 1626, describing a dinner for the French Ambassador attended by St. Antoine, notes: "This celebrated exponent of the equestrian art ... is on terms of familiarity with most of the nobility and gentry" (J. P. Hore, *History of Newmarket*, 1886, Volume I, pp.171–172).

The Tutbury Royal Stud of Charles I (r. 1625–1649): Running-Horse and Hobby Strains

Buckingham and Charles I

One of St. Antoine's most prominent pupils was Sir George Villiers, the future Duke of Buckingham (1592–1628). A letter written December 7, 1615, observed: "Every morning Sir George Villiers is a-horseback and taught to ride [by] St. Anthony, the Rider" (J. P. Hore, *History of Newmarket*, 1886, Volume I, p. 176). James I was enchanted with the handsome and accomplished rider Villiers who, as the Duke of Buckingham, came to be known as "The Grand Vizier" because of his political influence. When he was only 24 years old, Sir George Villiers was appointed Master of the Horse.

The Duke of Buckingham's love of dressage and his eminence as a rider was reflected in the bloodlines of the mares he selected for the Tutbury Royal stud and the stallions to which they were bred. Most of the mares in the 1624 inventory were bred to stallions with dressage bloodlines, to the Courser imported from Naples by Sir George Digby; the Spanish "Ginnets" (Jennets); and to "The French Horse" (Prior, *Royal Studs*, 1935, p. 48).

The description of the quarter-of-a-mile race at the city of Lincoln illustrates King James I's love of sprint racing. This enthusiasm was shared by his son, Charles I. After the death of his father, James I, Charles I retained the Duke of Buckingham as his Master of the Horse. On August 23, 1628, at 36 years old, the Duke was assassinated. He had taken an active interest in Tutbury, but his successor, the Duke of Hamilton, preferred a military career, neglecting Tutbury.

Newcastle and Charles I

Prince Charles first met William Cavendish, later the Earl, then Duke of Newcastle, when they were pupils at the school of horsemanship headed by the Chevalier de St. Antoine. A brilliant rider, Cavendish was 7 years

Right: Charles I by Van Dyck (1633).
Charles I and Newcastle studied together under the Chevalier de St. Antoine, instructor in dressage (since 1604) to the young noblemen. St. Antoine is shown in the portrait, holding Charles' helmet.

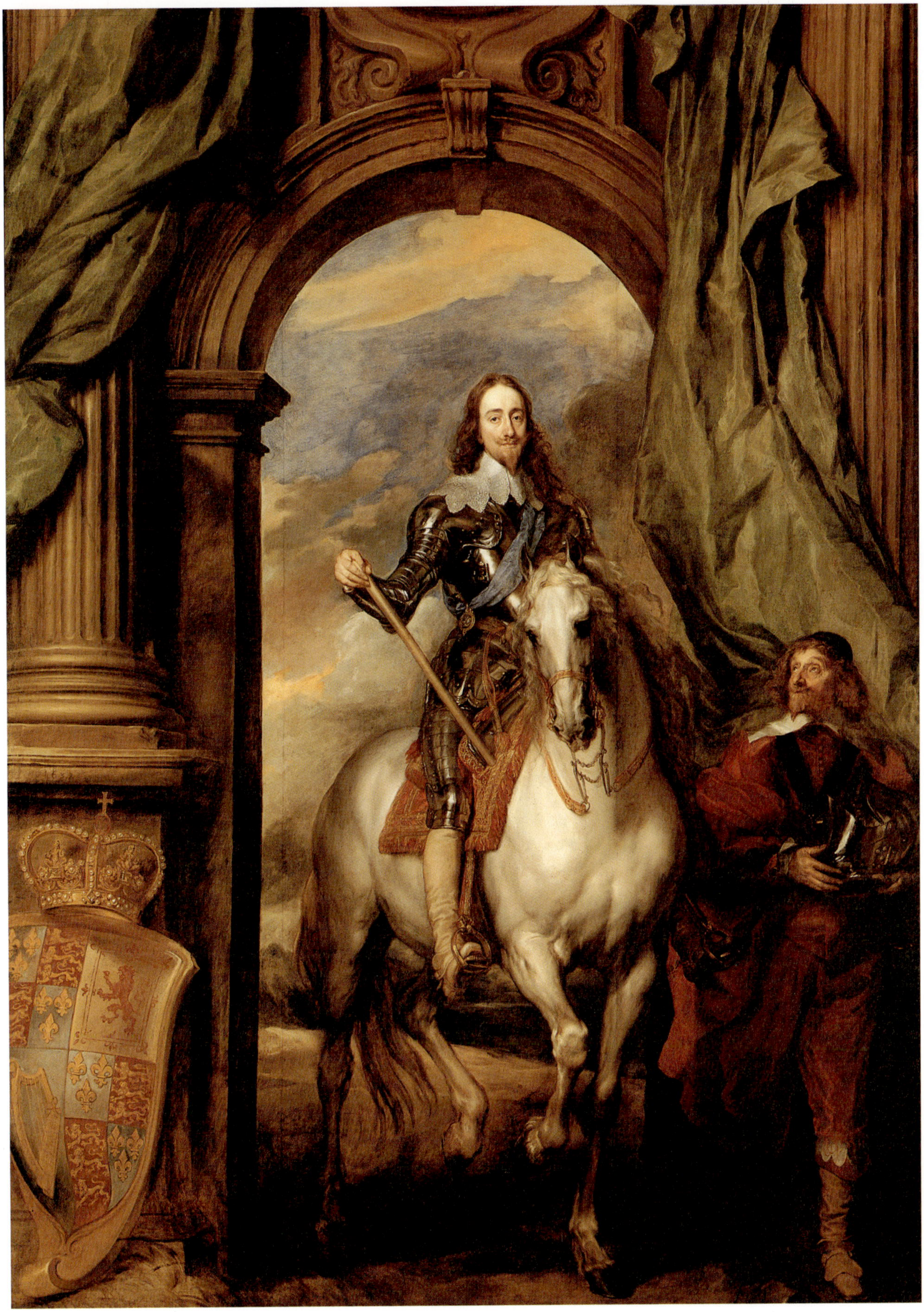

"An account of all his Ma^{ts} Mares and Colts within the race [stud] of Tutbury in Staffordshire, and with what Horses the breeding Mares were covered in this year, 1624" (The Royal Studs *by C.M. Prior, p. 40*).

older. Cavendish became a friend and trusted adviser to Charles I, who created him Duke of Newcastle on March 7, 1628. In 1638, King Charles I appointed Newcastle as Governor of the future King Charles II, then 8 years old.

Sir John Fenwick, Surveyor of the Tutbury Stud

Probably in 1628, with the encouragement of the Duke of Newcastle, Charles I appointed Sir John Fenwick as "Surveyor of Tutbury." Sir John was England's largest breeder of Running-Horse sprinters, with the most powerful racing stable. The main purpose of the appointment was to replace the dressage strain mares left in the Tutbury paddocks, which had been collected by Buckingham while Master of the Horse, with Running-Horse and Hobby strain mares. For the next 15 years or so, Fenwick was in full charge of Tutbury. Except for the gift in 1637 by the Emperor of Morocco of four Barb stallions to Charles I and records of a Parliament-ordered inventory in 1649, we have little documentary evidence as to Fenwick's term of office and his accomplishments.

In November 1647, Parliamentary forces on the Isle of Wight took custody of King Charles I as he was attempting to escape to France. Fourteen months later, on January 30, 1649, Charles I was executed by order of Parliament, which shortly thereafter sequestered the Tutbury stud, the personal property of the King.

The 1649 Tutbury Inventory

On July 27, 1649, an inventory of the 140 horses at Tutbury was drawn up by a Parliamentary committee (Prior, *Royal Studs*, 1935, p. 56). It contains no information as to pedigrees. Sir John Fenwick, Surveyor of the Race, who could have furnished full particulars of every horse, was not at Tutbury on inventory day. Still at Tutbury was Gregory Julian, "Yeoman of the Race" (stud groom), appointed during the reign of James I. Julian would not have greeted the inventory takers with enthusiasm; if they asked for pedigree information, it seems he did not supply it. The only clues as to pedigrees are contained in the names of the horses. These names are a mixed lot. Some of them appear to be the names of the horses themselves, while others appear to be the names of the sires of the horses.

The list includes the names of thirty-six brood mares of which twenty-four had foals at foot. There are six mares and four young horses whose names include the word "Newcastle." Apparently these represented gifts to or acquisitions by the King, from the Welbeck Stud of the Duke of Newcastle. We can be sure the racing stock acquired by the Tutbury stud carried Newcastle's best sprinting bloodlines; for his King, nothing less would do.

Two mares of special interest were "Carelton, one grisled mare with a little moone in her forehead with a horse foale 25 00 00" (25 pounds) and "Carelton, one black mare with a starre, 12 years old, 18 00 00" (18 pounds). Paragraphs from the 1607 and 1617 editions of Gervase Markham's *Cavalarice* refer to the 'Maister' Thomas Carlton's black

Hobby who beat "all the best Barbaries" at Salisbury. Evidently the Tutbury mares carried Carlton Hobby bloodlines. The evaluation of the first mare was the highest of all the twenty-four with foals.

Finally there is "Sorrell Fenwick. One sorrell mare with a blaze, 9 years old, with a mare foale, 18 00 00." We can be sure that as a gift to the King (and employer) this mare was one of Sir John Fenwick's choicest mares, and we also can be sure her pedigree included the speediest strains of Running-Horse bloodlines.

Barb strains in the Tutbury stud were represented by five mares and one younger horse whose names contained the word Morocco. In a letter written November 9, 1637, the Archbishop of Canterbury noted that an Ambassador from the Emperor of Morocco had just arrived with presents for the King, including four stallions. Presumably the six "Morocco" horses were descended from one or more of these stallions.

In 1650, the Tutbury racing stock was dispersed by sale or gift to prominent political figures and military officers, who would have had little interest in racing. The fences and stables were destroyed. By virtue of the random dispersal through patronage gifts or preferred sales, one of England's finest collections of sprinting strain bloodstock was broken up almost overnight.

The Breed Faces a Doubtful Future

In 1651, the future of the British racehorse breed seemed doubtful. In 1644, the Third Lord Fairfax, Cromwell's famous cavalry general, had captured and razed the Rutland family's North Yorkshire Helmsley estate, whose mares were the prime source of Hobby speed. In the same year, the Duke of Newcastle, forced into exile, had abandoned his Welbeck stud and its Running-Horse mares. In 1648, Fenwick's servant R. Watson, reported a tragedy, "that my master by the English and Scots [soldiers] is robbed and spoyled of all his breed of horses and mares, which loss with an addition of his beasts and spoyle of his house is valued to be above £3000." Fenwick died a decade later.

In 1650, the Tutbury Royal stud had been destroyed, its bloodstock widely scattered. The four principal studs of racing bloodstock were gone. Unless sufficient numbers of foundation mares from these four stud farms could be reassembled, the breed would be lost. A miracle was required.

CHAPTER VI

Perils and 'Saviors:' Foundation Mares and Their Breeders After 1650

The second half of the seventeenth century was the period in which the fragile Thoroughbred breed, battered during England's civil war, was miraculously restored. It is also the period of the earliest mares registered in the *General Stud Book*.

In 1791 James Weatherby, keeper of the Match Book at Newmarket and Proprietor of the *Racing Calendar*, published *An Introduction to a General Stud Book*, Volume I. This was an enormous success; with very few additions, it was printed nine times from 1791 to 1891 in five editions. *Speed and the Thoroughbred* is based on the 1891 edition.

Also based on this 1891 edition is *Breeding Racehorses by the Figure System* by the Australian turf scholar C. Bruce Lowe, published in 1895. For many hundreds of years, when selecting brood stock, British breeders had been influenced by stallions and tail male bloodlines. Bruce Lowe's was the first major work devoted to brood mares and tail female lines. It became the most influential book of the late nineteenth and early twentieth centuries.

Bruce Lowe produced a classification of tail female lines based on performance. He traced the tail female pedigrees of the winners of the five classic races for 3-year-olds (Epsom Derby, Oaks, 2000 Guineas, 1000 Guineas, and St. Leger) back to their earliest Stud Book ancestresses. There were 43 families which he called tap root mares, numbered in accordance with the total number of races won by each family. In his text the author gives special importance to tap root mares numbered one to fourteen. Number fifteen, the Royal Mare, Grey Whynot, discussed in Chapter IX also produced influential get and should be added to this list. The figure system is now obsolete, but the classification is generally accepted.

Perils

During the English civil wars (1642–1660) the British racehorse breed narrowly escaped destruction. Of the four principal studs whose mares had Hobby and/or Running-Horse bloodlines, Helmsley stud was captured by Oliver Cromwell's famous cavalry general, the Third Lord Fairfax in 1644, and the Welbeck stud had to be abandoned by the Duke of Newcastle, who was forced into exile in the same year. In 1648 Sir John Fenwick's Wallington stud was destroyed and its mares scattered by marauding soldiers. Finally, in 1650, Parliament gave away or sold at bargain prices the mares of the Tutbury stud, collected by Fenwick for King Charles I for over 20 years.

'Saviors'

Lord Fairfax

Ironically, the first person to reverse the trend in the misfortunes of the breed was the man who began them, Lord Fairfax, the breed's first 'savior.' His intervention contributed to the breed's revival. He had been wounded in 1644 during the siege of Helmsley Castle, and in 1651 Parliament gave him the entire Helmsley estate, including its livestock, "as a salve for an old wound." Although he lived at Nun Appleton and wrote a treatise on breeding cavalry horses, Fairfax made good use of the Helmsley bloodstock. He mated the Old Morocco Barb, probably a 1637 gift by the Emperor of Morocco to Charles I, to the foundation Hobby strain mare Old Bald Peg. She had been bred at Helmsley about 1635 by the Duchess of Buckingham. The Duchess was the only child and heiress of the Sixth Earl of Rutland, who gave her Helmsley in 1620 as part of her dowry when she married the First Duke. Old Bald Peg's daughter, the result of this mating, was the Old Morocco Mare (Bruce Lowe #6, *G.S.B.*, p. 14), the dam of Spanker, by the D'Arcy Yellow Turk. The *General Stud Book* (*G.S.B.*, p. 14) called Spanker "the best horse at Newmarket in Charles II's reign."

Constable Burton in the North Riding of Yorkshire, the seat of Sir Marmaduke Wyvill.

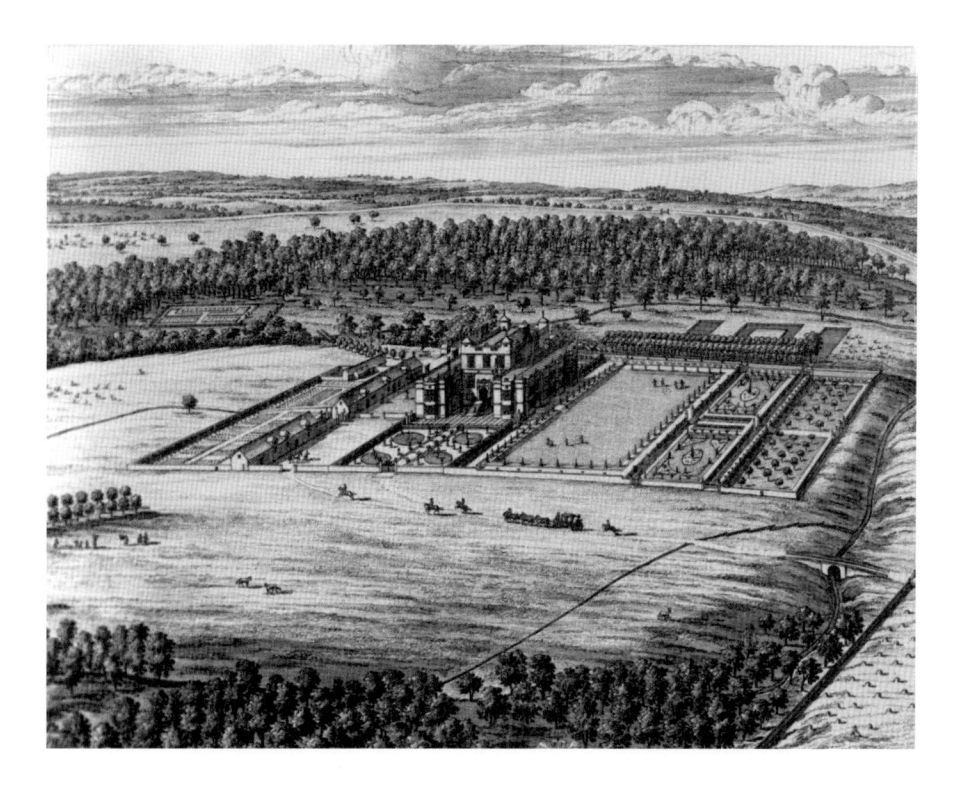

James D'Arcy the Elder of Sedbury

The second 'savior' of the Thoroughbred breed was James D'Arcy the Elder (1617–1673), the sixth son of the Seventh Lord D'Arcy de Knayth of Hornby Castle. D'Arcy had two careers, the first before the Restoration to the throne of Charles II in May 1660, the second after the Restoration until his own death in 1673.

In the mid-1640s, James D'Arcy the Elder married Isabel, daughter of Sir Marmaduke Wyvill (1591–1648) of Constable Burton Hall, member of a family long associated with the turf in Yorkshire. Isabel brought to her husband, as part of her dowry, the Sedbury estate, near Richmond, North Yorkshire, and perhaps some of the Wyvill bloodstock.

The D'Arcys, devout Roman Catholics and ardent supporters of the exiled King Charles II, were a family that loved racing. James D'Arcy apparently was well aware of the disaster to racehorse breeding inflicted by England's civil war. He must have believed that the Restoration of the King was imminent.

The Royal Mares

From happenings after Charles II regained his throne in 1660, we can trace D'Arcy's activities at Sedbury in the decade prior to the Restoration, following the dispersal of the Tutbury Royal stud bloodstock. To fulfill a contract, entered into in 1661, to supply the King annually with "12 extroynary good colts," D'Arcy would have to had already possessed 35 to 40 racing strain brood mares. Several D'Arcy mares and their daughters and granddaughters are registered as Royal Mares in the *General Stud Book*. D'Arcy had probably obtained some of the Tutbury Royal stud mares sold or given away by Parliament in 1650.

Cheny's 1743 *Royal Mare Falsehood*

In his 1743 *Racing Calendar* John Cheny said (mistakenly) that Charles II "sent abroad" for brood mares, a statement copied in *G.S.B.* (p. 388) and generally believed. However, the contract with D'Arcy fully supplied the King with racehorses. He had no need to send abroad for suitable mares (Prior, *Royal Studs*, 1935, p. 101).

John Fenwick

As the last Surveyor of the Tutbury stud, Sir John Fenwick would have been able to, and probably did, identify and study the breeding of the Tutbury mares, some of which were probably collected by D'Arcy after their dispersal from Tutbury. Fenwick also may have retrieved some of his own Wallington stud mares, dispersed when that stud was destroyed in 1648, mares which he might have sold to D'Arcy. Fenwick died in 1658.

That Sedbury continued to breed from Fenwick stock is recorded in the manuscript stud records of John Holles, the Duke of Newcastle, whose wife was a granddaughter of William Cavendish, the First Duke. These records were reproduced in Prior's *Early Records of the Thoroughbred Horse*, 1924, pp. 97–142. An entry dated January 25, 1703, reads: "Paid Mr. D'Arcy for 3 breed Mares 2 Dark sorrill Mares and a grey Mare 80 Guyneys." The sorrel mares were mother (foaled 1690) and daughter (foaled 1700). The older mare, known as Sorrill Whitenose, is described as "by Hoboy [Hautboy] her dam [c. 1680] extroynary fine out of Fenwick's breed."

That James D'Arcy the Younger greatly valued Fenwick's bloodlines is shown by an exchange of letters, also printed, in which D'Arcy tried to convince John Holles, the Duke of Newcastle and his wife, the Duchess, that he had been promised fillies out of the three mares. All he got from these letters however were indignant denials that Their Graces had ever made such a promise.

Perils and "Saviors": Foundation Mares and Their Breeders After 1650

Sedbury Hall, brought to James D'Arcy the Elder as a part of his wife's, Isabel Wyvill's, dowry in the mid-1640s. The greater part of the mansion was constructed in the early 1700s under the supervision of James D'Arcy the Younger.

James D'Arcy the Younger (1650–1731) of Sedbury, Created Lord D'Arcy of Navan (1721)

The agreement annually to supply the Royal Stables with "12 extroynary good colts" finally came to an end with the death of James D'Arcy the Elder in 1673. For the following 58 years, until his death in 1731, James D'Arcy the Younger dominated British racehorse breeding. In a petition to Queen Anne (r. 1702–1714), he boasted that he had "the best breed of horses in England," self-serving but undeniable. He had inherited the best foundation bloodlines and a fine stud farm in the Vale of Bedale in North Yorkshire, the center of race horse breeding.

D'Arcy made the most of these exceptional opportunities. Racing in the seventeenth century and much of the eighteenth century was a sport

James D'Arcy the Younger, painted c. 1680–1685 by Mary Beale (1632–1697). *D'arcy the Younger was created Lord D'Arcy of Navan 1721. From 1650 to 1660, his father, James D'Arcy the Elder, salvaged some of the dispersed mares of the Fenwick and Newcastle studs and the Tutbury Royal stud, saving the breed from extinction. D'Arcy the Younger inherited (1673) the Sedbury stud farm in the Vale of Bedale, Yorkshire, and its racing and breeding stock. Following a breeding plan based largely on the Royal Mares collected by his father, Master of the Studs to King Charles II, Lord D'Arcy built Sedbury into the era's largest (over one hundred brood mares) and most influential stud of racehorses.*

reserved for royalty, nobility and gentry. Even in the frontier colony of Virginia, this restriction was observed. On September 10, 1674, the Court of York County, Virginia, fined James Bullocke, a tailor, "100 pounds of tobacco and caske" for racing his mare, "it being contrary to law for a labourer to make a race" (A. Mackay-Smith, *The Colonial Quarter Race Horse*, 1983, p. 58).

Although he was a grandson of the Seventh Lord D'Arcy de Knayth, James the Younger, having no title, soon began to climb the ladder to nobility. In 1677 he was appointed a "Gentleman of His Majesty's Most

Honourable Chamber in Ordinary, to Wayt and Attend Upon His Majesty [Charles II] in Christmas Quarter." When he was 29 years old (1679), recognizing his increasing reputation as a horseman, the King issued an order to "admit James Darcy [sic], Esqr., to be Equerry to ourselfe in Ordinary of our Hunting Stables."

Petitions to King William III and Queen Anne

At the time of his death in 1673, James D'Arcy the Elder was owed £1000 for colts delivered to the King. James D'Arcy the Younger tried without success to collect his father's debt from King Charles II and King James II. When William became King (1689), he tried again. As an alternative to cash, D'Arcy suggested the following:

> Now your Royal Fleet [is] lying in the Mediterranean, that your Majesty would give order that good Barbary or Arabian horses be sent for, and that he may have six horses to supply the great number of breeding mares which your Petitioner hath.
>
> For in all England he cannot be furnished with good stallions but what are of the same kind [the same bloodlines as those at Sedbury] and the hazard of venturing eight hundred or a thousand pounds by the Merchants for such horses he cannot undergo, considering the great debts upon his estate.

The request for six stallions and reference to his "great number of breeding mares" suggest the size of the brood mare band at Sedbury then exceeded one hundred (eighteen mares per stallion), probably the largest collection of top-class mares in England. The Sedbury stud's need for stallions whose bloodlines would provide an outcross was due to Place's White Turk (*G.S.B.*, p. 388). At the time D'Arcy wrote this letter, the bloodlines of this great sire and his descendants had saturated the breeding stock of the North Yorkshire stud farms and "all England."

Apparently D'Arcy thought the Sultan of Turkey and the Emperor of Morocco might make diplomatic gifts of stallions to King William which could be put aboard British warships and shipped to Sedbury. Nothing came of this fanciful suggestion. It is interesting to note that D'Arcy considered the purchase of near-Eastern stallions from merchants a "hazard" and their prices of £800 or £1000 enormous.

Queen Anne was the fourth sovereign from whom James D'Arcy tried to collect the debt of £1000 owed his father for the colts delivered to Charles II. Queen Anne loved horse racing. She bred and campaigned her own horses and frequently went to race meetings. This time as payment, instead of cash or stallions, D'Arcy tried public office. When the Queen came to the throne (1702), D'Arcy presented a different petition. He suggested what Prior (*Royal Studs*, p. 96) called a "solatium," namely that Queen Anne appoint D'Arcy "First Equerry and Gentleman of the Horse to Her Majesty under Her Master of the Horse." This office was then held by a Mr. Ireton. D'Arcy hoped that "Her Majesty will be graciously pleased to let him succeed Mr. Ireton in case she orders him to be moved." There is no evidence that the Queen was pleased to do anything of the kind.

The reasons given by D'Arcy in support of his petition read: "He having the best breed of horses in England to supply Her Majesty—particularly if she sends any presents of horses to any foreign Princes." D'Arcy's statement that he had the "best breed of horses in England" was undoubtedly correct.

Attendance at Race Meetings

D'Arcy the Younger promoted and sold his horses not only at court, but also at race meetings which he frequently attended and where they raced many times. In 1684, D'Arcy wrote to his agent Christopher Croft at Whitton Castle, County Durham, instructing the latter to buy for him "a good pad, fine, strong and well going, such a beast as will mount me handsomely at a horse-race or public company." (Prior, *The Field*, Feb. 7, 1929, p. 214)

The *Sportsman's Dictionary* (London 1735) under the heading "Stallions" reads: "For the pad or ambling horse a Turkish sire and an English mare will naturally produce for the pace or amble..." D'Arcy's requirement for a "pad...as will mount me handsomely at a horse-race" needed racing blood, it then being the custom for the noblemen and gentlemen in attendance to view the races from the backs of their horses, which they galloped beside the front runners during the latter part of each race down to the finish line. Paintings and prints after Tillemans, Wootton and Seymour of racehorses nearing the finishing post at Newmarket show their supporters on horseback galloping beside the racing horses.

Lord D'Arcy of Navan

In 1721, when he was 71 years old, D'Arcy realized a lifelong ambition. He was created Lord D'Arcy of Navan in the Kingdom of Ireland, a distinction he enjoyed for the last 10 years of his life. This honor was conferred by George I on the recommendation of Sir Robert Walpole, importer of the Walpole Barb. Rival horsemen spread the jealous rumor that there was a connection between the horse and the peerage.

The *General Stud Book* tends to identify any horse bred from 1673 to 1731 at Sedbury as "bred by Lord D'Arcy." More correctly, horses bred between 1673 and 1720 should have been identified as bred by James D'Arcy the Younger, who only later became Lord D'Arcy.

The Huttons of Marske

Lord D'Arcy of Navan had four wives, but unfortunately did not sire a surviving male heir to carry on the Sedbury stud. Two daughters acquired Sedbury stud stock, however. His daughter Mary married William Jessop, a lawyer, who for £100, purchased from his father-in-law one of the best of the D'Arcy Royal Mares. In 1708, Jessop presented this mare to his principal client, John Holles, the Duke of Newcastle, whose wife was a granddaughter of the First Duke. The Jessops seem to have had little to do with the Sedbury bloodstock.

Lord D'Arcy's daughter Elizabeth, on the other hand, was a race horse breeder in her own right. While she was still a child (born 1706), her father evidently gave her one of his choice mares, Grey Royal (Bruce Lowe #11, *G.S.B.*, p. 10), carrying the best of the Sedbury foundation bloodlines. The mare's sire was the D'Arcy White Turk, her dam by the D'Arcy Yellow Turk, her granddam a Royal Mare. Grey Royal produced four fillies. The first (*G.S.B.*, p. 16) by Blunderbuss, went to her uncle Kitt (Christopher) D'Arcy, born 1653, her father's younger brother. The second, by Makeless, was sold to Captain Hartley. The third, by the Newcastle Turk, named Duchess and foaled in 1719, was sold to the Second Duke of Devonshire. All three fillies are bracketed in the *General Stud Book* under "Miss D'Arcy" as the breeder. It seems that Elizabeth could not bear to part with the fourth, (sire not recorded), who consequently is registered in the *G.S.B.* (p. 15) as Miss Betty D'Arcy's Pet Mare. When put to stud, these fillies made notable contributions to Thoroughbred bloodlines. In 1726, Elizabeth D'Arcy (1706–1739) married John Hutton, Jr. (1691–1768) of Marske, not far from Sedbury, member of a family long associated with North Yorkshire racehorse breeding.

D'Arcy breeding accounted for two Bruce Lowe numbers, #5 (Mr. Massey's Black Barb Mare) and #6 (Old Bald Peg), both Helmsley Hobby strain mares. Nine are Sedbury Running-Horse and Hobby strain mares, salvaged during the 1650s by James D'Arcy the Elder. Numbers 7, 11, 12, 13 and probably 15 are registered in the *General Stud*

Book as 'Royal Mares,' descended from Running-Horse and Hobby strain mares in the Tutbury Royal stud of Charles I. Number 2 is registered as "Mr. Burton's Natural Barb Mare" (*G.S.B.*, p. 4). In fact, this was a D'Arcy mare by a Barb stallion, standing at Constable Burton Hall, the seat of Sir Marmaduke Wyvill, father-in-law of James D'Arcy the Elder. As suggested by the *Stud Book*, this mare is the same as Bruce Lowe's number 3, known as the dam of the Dam of the Two True Blues (*G.S.B.*, p. 5). Of the first 15 Bruce Lowe mares, 11 were the property of the Helmsley and Sedbury North Yorkshire studs.

As noted in the previous chapter, the July 1649 inventory of the Tutbury stud was the source (through Prior) of the names of Running-Horse mares from the studs of the Duke of Newcastle and Sir John Fenwick, and Hobby mares from the Stud of 'Maister' Thomas Carleton.

The Layton Barb Mare, Bruce Lowe number 4, was a D'Arcy mare by a Barb stallion standing at Layton, a D'Arcy property. The Burton Barb and Layton Barb probably were sons of Helmsley's Old Morocco Barb, one of four presented in 1637 by the Emperor of Morocco to King Charles I. Number 8, the Bustler Mare (by Bustler), is the tail female ancestress of the Huttons' Marske 1750, sire of Eclipse and also of the important twentieth century sire, Bold Ruler 1954. On page 30 of his 1924 *Early Records of the Thoroughbred Horse*, C. M. Prior cites Cuthbert Routh's pedigree records which extend the tail female line of the number 8 mare back three generations to number 4, the D'Arcy Layton Barb Mare.

Three Generations of Huttons Bred Racehorses

The Huttons were breeders of one of the most illustrious tail female lines in the *General Stud Book* (*G.S.B.*, pp. 34, 395), the Bustler line. The first of the Hutton racehorse breeders was John Hutton, Sr. (1659–1730). In 1700, King William III gave him an imported stallion, thenceforward known as the Hutton Grey Barb (*G.S.B.*, p. 390). Second was John Hutton, Jr. (1691–1768), husband of Betty D'Arcy, breeder of Miss Betty D'Arcy's Pet Mare. Hutton Jr. turned over the Marske stud before 1750 to his son, John Hutton III (1730–1782). This last Hutton is credited as the breeder of Marske 1750 (descendant in tail female of the Bustler Mare by Bustler), sire of Eclipse 1764.

The breeder of Eclipse was the Duke of Cumberland (1721–1765). It is fortunate that he acquired Marske as a weanling from John Hutton III by trading for an Arabian stallion. It is also fortunate that George Stubbs was commissioned to paint Marske's portrait. This shows a dark brown

Perils and "Saviors": Foundation Mares and Their Breeders After 1650

Marske by George Stubbs (1724–1806).
In 1750, the Duke of Cumberland gave an Arabian stallion to John Hutton, III of Marske, grandson of Lord D'Arcy of Sedbury. In exchange, the Duke received a weanling named Marske 1750, a grandson of the Darley Arabian. Marske was the sire of Eclipse 1764. This painting was probably commissioned by the meat merchant, William Wildman, who quietly purchased Marske after he discovered Eclipse's extraordinary speed. The conformation of this horse shows a typical sprinter of the type bred by the D'Arcys at Sedbury from 1650 to 1731.

Marske Hall, home of the Hutton family. Three generations of Huttons bred racehorses.

horse, typical of the sprinters bred at Sedbury by Hutton's grandfather, Lord D'Arcy. Marske was a horse of great quality. He was compact, deep-bodied and short-legged. His beautiful head and high-set tail reflected the ancestry of his great-grandfather, the Darley Arabian. He was heavily muscled with somewhat straight [upright] shoulders—classic sprinter conformation. There are no crosses in his pedigree of the Godolphin Arabian, who contributed sloping shoulders to the Thoroughbred breed.

CHAPTER VII
IMPORTED 'PURE' ARABIANS: *GENERAL STUD BOOK* ENTRIES

The Fourth Part of the 1891 edition of the *General Stud Book* (pp. 388–394) lists 104 stallions, most of them with Barb, Turk or Arabian backgrounds. These are arranged more or less chronologically from the Markham Arabian 1616 to the Wellesley Arabian 1803. Most of the 25 stallions imported after 1750, with whom Mr. Weatherby, compiler of the first (1791) edition, was familiar but who had little lasting influence on the breed, are provided with dates. Virtually no importation or any other dates are provided as to origin for the 79 earlier stallions, even for such important and well documented sires as Place's White Turk 1657, Curwen's Bay Barb 1698, the Darley Arabian 1704 and the Godolphin Arabian 1730.

In 1689, in his petition to King William III, James D'Arcy of Sedbury complained: "The hazard of venturing eight hundred or a thousand pounds by the [eastern] Merchants he cannot undergo." It is not surprising that these eastern importations were few in number:

- Buyers could not see their new stallions until they were delivered.
- There were no performance records of any kind.
- Top class stallions were difficult to procure, even through diplomatic channels.
- Prices were elevated by competition from Polish, Hungarian and Russian racehorse purchasers.

- Transportation overland from Constantinople, or by sea from Smyrna, was hazardous and a further expense.

The Markham Arabian

The first and earliest horse listed as an Arabian in the Fourth Part of the *General Stud Book* is the Markham Arabian (p. 388). The *G.S.B.* notes he was "said (but with little probability) to have been the first of the breed ever seen in England." It also quotes in part a description of the horse written by the Duke of Newcastle in his *A New Method and Extraordinary Invention to Dress Horses* (London, 1667, p. 73). The full text reads:

> I never saw any but one of These Horses, which Mr. John Markham, a Merchant, brought Over and said, He was a Right Arabian: He was a Bay, but a Little Horse, and no Rarity for Shape; for I have seen Many English Horses farr Finer. Mr. Markham Sold him to KING JAMES for Five Hundred Pounds; and being Trained up for a Course, when he came to Run, every Horse beat him.

Even though Newcastle had traveled extensively in Europe, he had never before seen an Arabian.

On December 20, 1616, a payment from the Exchequer for receipts of James I of £154 was made to Master Markham for an Arabian horse, and a further payment of £11 to the man who delivered him. Earlier in his chapter, *The Arabian Horse*, Newcastle wrote: "I have been Told by many Gentlemen of Credit, and by Many-many Merchants, That the Price of Right Arabians is, One Thousand, Two Thousand, and Three Thousand Pounds a Horse (an Intolerable, and an Incredible Price)." The discrepancy in price probably indicates that the Markham Arabian was British-bred. There are no documentary records of any descendants.

Aleppo

The Arabian desert in Syria was part of the Ottoman (Turkish) Empire whose capital was Constantinople. The nomadic Mesopotamian and desert sheiks traded at Aleppo, the western terminus of the Asian caravan routes. During the seventeenth and early eighteenth centuries the word Arabian was used to designate a horse bred by the nomadic tribes of the desert regions east and south of Aleppo—Bedouins and others.

Imported 'Pure' Arabians: General Stud Book *Entries*

The city and castle of Aleppo in Syria, an engraving by A. Drummond.
Situated at the top of the divide between the Euphrates Valley and the Mediterranean, Aleppo was the great entrepôt of trade between Europe and the East. More than 1,000 acres of the level plain at the foot of the castle were covered with small shops offering goods of all descriptions. The Bedouin and other sheiks who traded at Aleppo occasionally offered an Arabian stallion for sale, but export was forbidden.

When Aleppo and its market were seized by the Turks in 1516, the Turkish Sultans were so overwhelmed by the beauty of these tribal horses, which they considered national treasures, that their export was strictly forbidden.

Aleppo, strategically situated on the mountain divide between the headwaters of the Euphrates River Valley to the south, and less than a hundred miles uphill from the harbor of Antioch, in the far eastern Mediterranean, was the entrepôt of trade between the Mediterranean and the lands to the east. Its twelfth century fortress dominates the plain and the city at its foot, famous for the great covered bazaar with myriad small shops offering goods of every description. The riches of India, China and the Spice Islands came by ship to the head of the Persian Gulf. From there the goods were trans-shipped by camel caravan, joined by wares from Tehran, Baghdad and Damascus, across Persia to Aleppo. Before the sixth century, there was no native breed of horses on the Arabian peninsula. It is conjectured that tolls exacted from the caravans for safe passage through the desert by the Bedouin sheiks enabled them to acquire the finest central Asian mares which founded the Arabian breed.

Competition among the Arab tribes for grazing areas and personal animosities led to tribal raids and open wars. Horses were used by the desert Bedouins of Arabia for warfare, not organized racing. (Horses described as Arabian, bred by more settled tribes farther to the north, however, *were* bred for racing.) When planning a raid, the desert warriors rode camels, leading horses to a hiding place, near the enemy tents. Since stallions might reveal their presence by neighing, only mares were used. Under cover of darkness, warriors swooped down on their sleeping foes, killing or wounding a few, gathering what booty they could carry, and setting off at the gallop on their mares for their own tents, perhaps 70 or 80 miles away. Their surviving victims were soon mounted and in hot pursuit. Over centuries, using mares selected for superiority in these chases of several hours, the Bedouins produced a breed of war horses with great distance capacity, spirit, stamina and quality, but with little 4-mile course speed. The export of both desert *and* racing Arabian horses was forbidden.

Arabian blood today dominates endurance riding competitions (100 miles, 24 hours). In 1982, endurance riding became an international sport when approved by the F.E.I. (Federation Equestre Internationale).

Turkey Merchants

Throughout the sixteenth, seventeenth and eighteenth centuries, Englishmen established themselves as merchants at trading centers in the Ottoman Empire, notably at Constantinople, Smyrna and Aleppo. Known to Elizabethans as *Turkey Merchants*, many made large fortunes. During the decades 1700–1720, two turkey merchants managed to export three "pure" Arabian stallions to clients in England.

Imported 'Pure' Arabians General Stud Book Entries

The Darley Arabian from an engraving after John Wootton.
Foaled in 1700, he was bought in 1704 by Thomas Darley, who shipped him out of Smyrna and sent him to his brother, Richard, at Aldby, Yorkshire. He stood until about 1730 at Aldby, latterly the property of John Brewster Darley. One of his sons was Bulle Rocke, the first Thoroughbred to go to America (1730). His two sons out of Edward Leedes' Betty Leedes (c. 1705) line bred to Old Bald Peg, dam of the Old Morocco Mare (G.S.B., p. 14) were Devonshire (Flying) Childers 1715 and Bartlett's (Bleeding) Childers 1716. The latter carried on the Darley Arabian tail male line through Eclipse 1764 to most Thoroughbred horses today. These two sons spread throughout the breed their sire's spirited temperament and exceptional beauty.

Darley's Arabian

A stallion who begot a major current in the Thoroughbred bloodlines mainstream was the Darley Arabian, foaled in 1700, from whom about 90% of all present-day Thoroughbreds descend in tail male. The entry in the *General Stud Book* reads (p. 391):

> Darley's Arabian, probably a Turk or Syrian horse, was brought over from Smyrna by [Mr. Thomas Darley] a brother of Mr. Darley, of Yorkshire, who being an agent in merchandise abroad, became member of a hunting club, by which means he acquired interest to procure this horse. He was sire of Childers, and also got Almanzor, a very good horse, and Aleppo his brother; also a white-legged horse of the Duke of Somerset's, full brother to Almanzor, and thought to be as good, but, meeting with an accident, he never ran in public; Cupid and Brisk, good horses; Dædalus, a very fleet horse; Dart, Skipjack, and Manica, good Plate horses, though out of bad mares; Lord Lonsdale's Mare in very good form, and Lord Tracy's Mare in a good one for Plates. He covered very few mares except Mr. Darley's, who had very few well bred besides Almanzor's dam.

Thomas Darley was a 'turkey merchant,' established at Aleppo. At the request of his father, Richard Darley, who lived at Aldby Park, Buttercrambe, near York, he procured a dark bay Arabian stallion of exquisite quality. According to the inscription on a life-sized portrait at Aldby Park, the horse was 15 hands high. It is said that Thomas Darley's membership in an Aleppo hunting club helped him fulfill his father's commission "for a tolerable sum." In January, 1704, the dark bay stallion was put on board the ship *Ipswich* at Smyrna, William Watkin, Captain. Later that spring, he arrived at Aldby Park. Here the Darley Arabian lived for the rest of his life, the property of three generations of the Darley family. Although on arrival he was only 4 years old, no attempt was made to race the Darley Arabian. We do not know the date of his death.

Leonard Childers of Carr House, Doncaster, in 1714, sent his mare Betty Leedes to Aldby Park to be bred to Darley's Arabian. Betty Leedes, foaled c. 1705, was bred by Edward Leedes (d. 1703) or his son Anthony, of North Milford near Tadcaster, Yorkshire. Her sire was Old Careless, foaled 1692 (by Spanker). Old Careless was the best racehorse at Newmarket in 1698. Bred by Mr. Leedes, he was purchased and raced by the Fifth Lord Wharton. In 1699 or 1700, he was leased as a stallion to Mr. Leedes, a leading market breeder. Mr. Leedes' foundation mares were two daughters of Helmsley's Old Morocco Mare (*G.S.B.*, p. 14).

Betty Leedes had in her pedigree three crosses of Helmsley's Old Bald Peg (Bruce Lowe # 6, *G.S.B.*, p. 14), the principal foundation mare of Hobby strain speed. From her union with Darley's Arabian Betty Leedes produced Flying (Devonshire) Childers 1715. She was bred back to Darley's Arabian that year and foaled Bartlett's (Bleeding) Childers 1716, the grandsire of Eclipse.

The Darley Arabian at Stud

Except for full brothers Devonshire (Flying) Childers and Bartlett's (Bleeding) Childers, the record of the Darley Arabian's offspring as a sire of racehorses was not impressive. It was in 1721 and 1722 that Flying Childers' extraordinary speed astonished British breeders and racegoers. One would suppose that breeders would accordingly send their best mares to the imported sire of "the fleetest horse that was ever trained in this or any other country" (*G.S.B.*, p. 389). Instead they sent their mares to the sons, Flying Childers and the unraced Bleeding Childers. British breeders were beginning to realize that winners must have speed strains in their pedigrees, and that it was the concentration of such strains in the pedigree of the dam, Betty Leedes, which produced the dazzling speed of Flying Childers and the remarkable stud career of Bleeding Childers.

Through these two sons and their get, the Darley Arabian spread throughout the entire breed his high spirits, his stamina, his exquisite beauty, his broad and concave forehead, his luminous eyes, his small ears and muzzle and his high-set tail. The Darley Arabian transformed the Thoroughbred into a beautiful horse. He will always be famous as the tail male progenitor of most present-day Thoroughbreds.

The Oxford Dun Arabian 1715

The other 'turkey merchant' was Nathaniel Harley. In 1686 he settled in Aleppo, where he lived for the rest of his life. He died in 1720. Nathaniel Harley's principal horse client was his nephew, Edward, Lord Harley, in 1724 created Earl of Oxford. In 1713 Lord Harley married Henrietta Cavendish Holles, great-granddaughter of the First Duke of Newcastle, thereby acquiring Welbeck Abbey in Staffordshire, together with its magnificent stables and covered riding school. For his invaluable volume, *Early Records of the Thoroughbred Horse*, 1924, C. M. Prior was able to examine the extensive family archives preserved intact at Welbeck, through the courtesy of the Duke of Portland, a direct descendant of the Duke of Newcastle.

Nathaniel Harley sent two Arabian stallions, who are both registered in the *General Stud Book*, to his nephew. The 1704 shipment of the Darley Arabian may have escaped the notice of the authorities, but when Nathaniel Harley sent the Oxford Dun Arabian (*G.S.B.*, pp 10, 389) from Aleppo to Welbeck Abbey in 1715, it was a difficult shipment. Prior (*Early Records of the Thoroughbred Horse*, 1924, p. 141) quotes a letter written February 15, 1714/15 [sic] by Harley to his brother, auditor Edward Harley:

> Three Expresses have been sent after him [the Oxford Dun Arabian], and all the passes of the Mountains between this [place, Aleppo] and Scanderone ordered to be watched, and ye marine strictly guarded to prevent his being ship'd off... I believe few such Horses have ever come to England... I've had so much trouble, Expence and difficulty at first to procure, afterwards to keep and now to send him away, that I think him above any price that can be offered, and am so little of a Merchant that I would not have him sold even tho' a Thousand Pounds should be bid for him.

The letter states that the bloodlines of the best Arabians stallions were considered national treasures, not only by the Arab sheiks who bred them, but also by the Turkish government. It also establishes the influence of Nathaniel Harley, a foreign merchant, who managed to hide the horse, by land and sea, from the Turkish officials, despite their extraordinary efforts to prevent the export of one horse.

On December 4, 1716, Lord Harley wrote his uncle from Welbeck:

> The fine Dun Horse wch you sent over and came safe about a year & a half ago is under my care & is very well, and is thought by all that have seen him to be the finest Horse that ever came over. (Prior, *Early Records*, 1924, pp. 141, 142)

According to other Welbeck documents, the Dun Arabian was bred about 1717–1718 to D'Arcy's Royal Sorrill Mare (not in the *General Stud Book*), foaled about 1703, by Wastell's Turk, a daughter of D'Arcy's Blacklegs (Bruce Lowe #7, *G.S.B.*, p. 16). The produce of this mating was a mare whose daughter, Miss Slamenkin (*G.S.B.*, p. 137), carried on the Bruce Lowe #7 tail female line. It includes such nineteenth century notable sires as West Australian, Flying Fox and Persimmon.

Lack of Arabian Speed

The Bloody Shouldered Arabian

Imported sires arrived in England without performance records. They were mostly mature horses. Their new owners did not race them. In consequence, the only way to assess the value of their bloodlines as sources of speed was through the speed of their offspring.

There is, however, one documented instance of an imported Arabian sire who was tried and found lacking in speed. This was a grey horse with a red mark on his shoulder, the Bloody Shouldered Arabian, sent to Welbeck Abbey during the winter of 1719–1720 by Nathaniel Harley.

Nathaniel Harley was not only an authority on Arabians, he was also an honest merchant. In the winter of 1719–1720, the Oxford Bloody Shouldered Arabian was shipped to his purchaser, Edward, Lord Harley. At the time of the shipment, the horse was 6 years old, but Nathaniel Harley had purchased him at two. During the 4 year interval, he had been tried for speed, presumably over 4-mile distances. In the letter of transmittal, to his nephew (January 6, 1720), Nathaniel Harley described the Bloody Shouldered Arabian: "He is of great spirit but no great speed, wou'd soon learn anything in the manage [dressage arena]." Like other imported Arabians, the Bloody Shouldered Arabian had spirit, quality

The Bloody Shouldered Arabian.

Exported with him went the story that his dam was the mount of a badly wounded highwayman trying to escape. Weakened by loss of blood, he leant forward over the shoulder of the mare, then in foal. His blood stains on her shoulder were said to have been inherited by her colt.

and stamina, but not speed (Prior, *Early Records of the Thoroughbred Horse*, 1924, p. 142). The *General Stud Book* (p. 391) notes only that he was the sire of the Bolton Sweepstakes winner, foaled 1722 (*G.S.B.*, p. 11).

The explanation for the red mark on the horse's shoulder, given to Nathaniel Harley, reads:

> The owner, he told me, of the mare that brought [produced] this colt was a Robber on the Road, and being much wounded he leant over his mare's neck, and his blood ran down upon her shoulder, and she being then with foal of this colt, he had the mark on his shoulder. (Prior, *Early Records*, 1924, p. 142)

The Bloody Shouldered Arabian sired a few useful racehorses, but his reputation came from his spectacular shoulder marking, not from his progeny. John Wootton, the most popular sporting artist of that day, painted only one portrait each of the Darley and Godolphin Arabians, but nine or ten of the Bloody Shouldered Arabian (Prior, *Royal Studs*, 1935, p. 136). Such was his reputation that in 1729 when he was 16 years old, the Sixth Duke of Somerset gave Lord Oxford "one hundred broad pieces" (£115) for the horse and moved him to Petworth, his stud farm in Sussex.

Daniel Defoe, Wm. Taplin & C. M. Prior

Daniel Defoe, author of *Robinson Crusoe*, in 1724–1726, published *A Tour Through the Whole Island of Great Britain*. In North Yorkshire, he saw several imported stallions. He wrote:

> They do indeed breed very fine horses here, and perhaps some of the best in the world, for let foreigners boast what they will of barbs and Turkish horses, and, as we know five hundred pounds has been given for a horse brought out of Turkey, and of the Spanish jennets from Cordova, for which also an extravagant price has been given, I do believe that some of the gallopers of this country [North Yorkshire], and of the bishopric of Durham, which joins to it, will outdo for speed and strength the swiftest horse that was ever bred in Turkey, or Barbary, taken them all together.

When he published his two volume *Sporting Dictionary* in 1803, William Taplin had sufficient perspective to assess the developments in racing and breeding which had taken place since 1700, as well as personal acquaintance with many of the owners and breeders, together with close observations of their horses. In discussing the use of imported Arabian sires (Vol. II, p. 208) he wrote:

> After having crossed the blood in all possible directions, numerous experiments were made (and for large sums) in bringing the different crosses to the post in opposition to each other; when, after every possible refinement, and every judicious exertion, to ascertain the superiority of the ARABIAN BLOOD, it was at length most clearly proved, that the more they bred *in and in* with the foreign horses and mares, the more they... became gradually SLOWER, and longer upon the ground, and *farther* they had to *run*. This discovery having been made... about the year 1760, the rage for Arabian extractions has been gradually upon the decline with the sporting aggregate from that period to the present time [1803].

C. M. Prior wrote:

> If the Arab can be said to have obtained the zenith of its success by about 1750, its decline in popularity must have set in with great rapidity, for by 1782 those 'by Arabians' were actually allowed 3 lb. [less rider weight] in the Cumberland Stakes at the First Spring Meeting at Newmarket. (*History of the Racing Calendar*, 1926, p. 123)

That the use of Arabian strains for a few generations slowed the speed of the earlier British-bred Thoroughbred is a conclusion diametrically opposed to ideas long held by the general public and asserted by many turf writers. Taplin's contemporary testimony, that leading British breeders of the eighteenth century were well aware of these developments, is of special importance and significance.

CHAPTER VIII

A Century of King's Plates, 1665-1780

Four-Mile Heat Racing

Charles II (1630–1685) was the most accomplished horseman of all the British Monarchs. He was an exceptional natural athlete. His instructor was the Duke of Newcastle (1593–1676), author of two famous books on dressage (1657, 1667), a bold rider to hounds and a successful rider of sprint races won by his home-bred Running-Horses. Newcastle was appointed "Governor" of Charles, then Prince of Wales, when the latter was 8 years old (1638).

In the biography of her husband (*The Life of William Cavendish, Duke of Newcastle* by Margaret, Duchess of Newcastle, 1667), the Duchess wrote:

> My Lord [the Duke] had the honour to set him [Charles II] first upon a horse of manege; where his capabilities were such, that being but ten years of age [1640], he would ride leaping horses, and such as would overthrow others, and manage them with the greatest skill and dexterity.

In his book, *A New Method and Extraordinary Invention to Dress Horses* (London, 1667), the Duke said of the young Prince that "he made my horses go better than any Italian or French riders (who had often rid them) could do." His opinion, written in 1667, reads:

> The King is not only the handsomest and most comely horseman in the world, but as knowing and understanding in the art as any man; and no man makes a horse go better than His Majesty, the first time that ever he came upon their backs, which is the height of the art.

From the ninth century to the present, British royalty has given constant support to horse racing. Charles II maintained a large stable of racehorses, campaigned them regularly in match races and plate races, and, on several occasions, rode the winners himself. This addiction was undoubtedly encouraged by the King's boyhood 'Governor,' the Duke of Newcastle, who wrote of his affection for the sport of Running-Horses, and his many hundreds of races ridden.

Charles I had lost his Kingdom and his head (1649) largely because of the superiority of Cromwell's cavalry. In 1660, when restored to the throne, Charles II fully realized the importance of strengthening his cavalry, both men and mounts. The improvement of the British racehorse would gratify the King's first love; to increase the supply of cavalry-type horses would fulfill his duty to defend Britain. Five years after he was restored to the throne, Charles II decided to take action. He was well aware that the center of racehorse breeding was in North Yorkshire's Vale of Bedale. Here was the Sedbury stud of James D'Arcy, whom, in 1660, he had appointed his Master of the Stud. Nearby was the Helmsley Castle stud, owned by the King's friend since childhood, the Second Duke of Buckingham. Adjacent to the Vale of Bedale was the Vale of Cleveland where for centuries farmers had produced a closely allied breed of general utility horses, then known as Chapman Horses, and since 1800 as Cleveland Bays.

It would have been logical to support both breeds individually. Instead the King apparently decided to create a new dual purpose breed, to have his cake and also eat it. Charles II believed, through royal patronage, he could induce breeders to produce a strain which would provide racing of high caliber, and also stallions suitable to sire cavalry mounts and horses of similar type. He further believed that these goals could be achieved through long distance races for horses carrying high weights.

From about 1512 to the outbreak of the English civil war in 1642, the annual short course sprint races organized by the major towns and cities attracted by far the largest crowds. William Camden's description in his 1586 *Britannia* reads in part:

> The Forest of Galteresse [the course near York] is famous for a yearly horse race...It is hardly credible how great a resort of people there is to these races from all parts and what great wagers are laid.

Cavalry horses had to gallop distances far longer than a quarter of a mile. In order to prevent the revival of municipal sprint races and to attract even greater crowds, the King devised an ingenious new set of "Articles" (racing conditions). The 8- to 12-mile gallops characteristic of match races between Hunting-Horses were replaced by three or four heats of 4 miles each. These new races, instituted in 1665, were called the King's Plates. So that the crowds could see and enjoy every part of the racing, a "New Round-Heate" course was constructed at Newmarket, which Charles II had made the center of British racing.

The King's Plate "Articles" can be summarized as follows: The race was to be run over Newmarket's "New Round-Heate" course. Every rider was to "weigh twelve stone, fourteen pounds to the stone [168 lbs.] besides saddle and bridle." Every horse was to run "the new round course three times over. Every horse shall have half an hour's time to rub between each heat." The Plate was to be awarded to the horse winning all three heats, but failing this, the heat winners were to run a fourth heat over "The Course," after an interval of an hour and a half. The race was open to any horse, mare, stallion or gelding. Horses and riders were summoned to the post by the clerk of the race "by the signal of a drum [or] trumpet, setting up an hour glass for that purpose." Other clauses dealt with foul riding, moneys to be subscribed and racing officials.

The length of the "New Round-Heate" course was actually 3 miles, 6 furlongs, 93 yards, while the length of "The Course" was 4 miles, 1 furlong, 138 yards. If the three heats were not won by the same horse (which frequently happened) and the winners of the different heats were raced over "The Course," the total distance covered was approximately *16 miles* with 168 lbs. of jockey plus saddle!

There is an old race track saying that "weight will stop a freight train." Horses that averaged about 14:2 hands carrying 168 pounds, racing over 12 to 16 miles, did not gallop very fast! In September 1699, the *London Gazette* noted the conditions of the King's Plates were designed to produce "strong and useful horses." *Nothing was said about speed.*

Charles II at Newmarket

Charles II demonstrated his support of the King's Plates not only in silver but also in person. Writing from Newmarket on March 24, 1675, Sir Robert Carr reported: "Yesterday, His majestie Rode himself three heats and a course and won the Plate, all fower [four] were hard and nere [close] run, and I doe assure you the King won by good Horseman-Ship." (J. P. Hore, *History of Newmarket*, 1886, Vol, II, p. 326)

To ride 16 miles in one afternoon in four races on the same horse and win, was a remarkable performance for the 45 year old Monarch who loved to "eat, drink and be merry." No doubt previous experience was a factor. He won the October Newmarket Plate in 1673, beating his natural son, the Duke of Monmouth, Mr. Elliott, and Mr. Thomas Thynne.

Preferring Newmarket over Whitehall and the affairs of state, the King began his day there with an early morning walk at so brisk a pace that courtiers asking for favors or presenting petitions were left far behind. Mounting his famous grey hack, Old Rowley, Charles then rode to watch the gallops on Newmarket Heath. Next came cock fights, the afternoon races, at night the gaming tables, and then the company of Louise de Kerouaille, Duchess of Portsmouth, Nell Gwynn and other beauties.

Hunting-Horse Foundation Bloodlines

During the sixteenth and seventeenth centuries, the strains with staying bloodlines, the horses that ran in long-distance match races, were known as "Hunting-Horses." In his 1593 treatise on training racehorses, entitled *Discource of Horsemanshippe* (see Chapter III), Gervase Markham makes the distinction between the fast sprinter known as the Running-Horse ("His Tryall he must dispatch in a moment") and the long distance stayer known as the Hunting-Horse ("His Tryall-of long and wearie toyle").

Hunting was restricted to the nobility and gentry. Red deer (the stag, the hart) were the property of the Crown; roebuck (the buck) were the property of the great landowners. The horses that were used to follow the stag hounds and buck hounds were stayers over distance and therefore suitable to run in distance races. At the major race courses, distance racing for Hunting-Horses ridden by their owners and their friends was a sport for the nobility and gentry, not for the "lower orders." A clause in the "Articles" specified that "no groom or serving man" shall be allowed to ride.

A Century of King's Plates, 1665–1780

A match race over the long course at Newmarket, an engraving by Peter Tillemans.

On October 16, 1665, when Charles II founded the series of King's Plates at Newmarket, he was able initially to base his new departure in British racing on a long-established pool of staying bloodlines, and on the numbers of Hunting-Horses and distance races. This combination gave the scheme an auspicious start.

King's Plate Performance Standards

The performance standards of the King's Plates were fully accepted by British breeders. King's Plates were established at other courses until there were over twenty in England and Scotland. The winning of a King's Plate assured the horse's reputation for success, on the race course and when beginning a career at stud. The prestige of these plates is indicated by the racing career of unbeaten Eclipse 1764, foremost in the eighteenth century both as a racehorse and sire. In 1769 and 1770, as a 5- and

Speed and the Thoroughbred

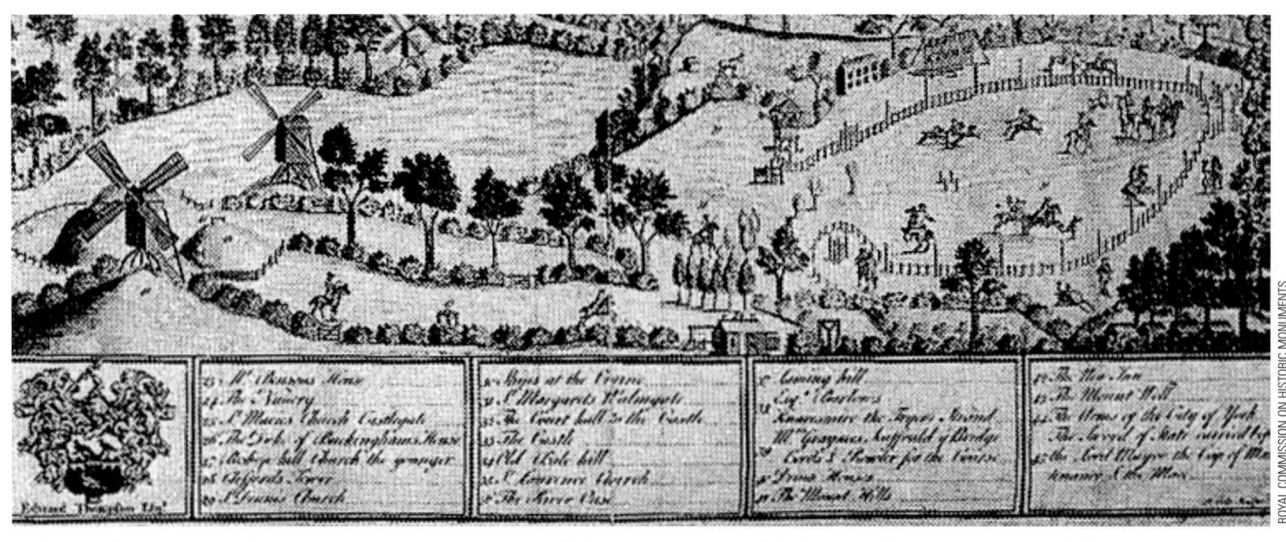

The newly-constructed Navesmire race course. Detail from engraving of "The South West Prospect of the Ancient City of York wth the Platt-form of Knavesmire, Wherein His Majesty King George the Second's Hundred Guineas was Run for Aug ye 16,th Anno Domini 1731." This was a race for 6-year-olds carrying 12 stone (168 pounds), two 4-mile heats, both won by Lord Lonsdale's Monkey.

Monkey 1725 by James Seymour (1702–1752), watercolor (gouache), 1731.
Monkey, by the Earl of Lonsdale's Bay Arabian out of a Curwen's Bay Barb mare, his grand-dam by the Byerley Turk. Monkey was imported to Virginia by Nathaniel Harrison, where he stood at stud from 1737 to 1749 at Harrison's Brandon plantation on the lower James River. He continued his career at stud in the Roanoke River Valley until his death in 1754.

6-year-old, Eclipse started in and won eighteen races. Ten of these were King's Plates, run at Newmarket (2 races), Winchester, Canterbury, Lewes, Litchfield, Guildford, Nottingham, York and Lincoln. Often only two winning heats were specified instead of three.

Breeding for Speed Replaces Breeding for Stamina

King's Plates dominated British racing for 100 years. The runaway victories of unbeaten Flying Childers 1715 and Regulus 1739 brought home to breeders the realization that it is not stamina that wins races, but speed. This led to a mounting demand for lower weights, shorter distances, fewer heats and younger horses. In the 1751 King's Plates run at Newmarket, King George II tried to stem the tide toward lower weights and shorter distances by ordering 6-year-olds to carry 12 stone (168 lbs), but limiting the permissible distance to four heats of 3 miles, 6 furlongs, 93 yards each (Pond's *Racing Calendar*, 1751).

The end of the era was foreshadowed by the institution of our present-day classic single heat races for 3-year-olds, in 1776 the Doncaster St. Leger at 2 miles, and in 1779 and 1780 the Epsom Oaks and the Epsom Derby, at a mile and a half. (Diomed, sire of unbeaten Virginia-bred Sir Archy and great grand-sire of Lexington 1850, won the first Epsom Derby.) Charles II's hope—to have his cake and eat it too—proved to be elusive. Breeders were interested in horses that could win races, not in producing "strong and useful horses."

Although they failed to achieve the primary objective of Charles II, the "Articles Ordered by His Majestie to be Observed" in the initial King's Plate, run on Newmarket Heath, October 16, 1665, and signed by the King, started a chain of events which brought about major changes in breeding and racing in the ensuing 100 years. These changes were:

- The King's Plates gave racing the prestige of Royal patronage, from Charles II through George III (7 Monarchs).
- The frequent presence of Charles II at Newmarket made it the world's center of racing, a distinction it enjoyed from the seventeenth to the beginning of the twentieth century, when exports of Thoroughbreds became widespread to the United States, Canada, France, Hungary, Australia, Japan and other countries.
- The Articles supplied a national performance standard (4-mile heats) for racing, and a set of rules as to how racing should be conducted, both previously lacking. Racing thereafter was conducted in an organized fashion.

Speed and the Thoroughbred

Diomed attributed to George Stubbs (1724–1806), oil on canvas.
Diomed was the winner of the first Epsom Derby in 1780. After an undistinguished career at stud, he was sent to Virginia and bred in 1804 to imported Castianira. The mating produced the unbeaten Sir Archy, known as the "Godolphin Arabian of America." Sir Archy was the grand-sire of Lexington 1850.

- Observance of these Articles provided the base for a tremendous expansion of British racing. This was evidenced by the annual publication of *Racing Calendars* containing the year's complete racing results, beginning with John Cheny's Calendar of 1727. This was followed in 1791 by James Weatherby's *Introduction to a General Stud Book*.
- The performance tests set forth in the Articles were fully accepted by breeders, and became the acknowledged Thoroughbred standard of excellence. At the front of each racing calendar appeared a list of the year's King's Plate winners.
- The longer distance racing of the Articles led to the abandonment of sprint racing and the extinction of the special short speed strains, the sixteenth and early seventeenth century Running-Horses, and the English and Irish Hobbies. Fortunately these sprinting speed strains were carried on in female bloodlines.
- Under the terms of the Articles, a race meeting would include both the sixteenth and seventeenth century longer distance match races (c. 6 to 12 miles) between two or three "Hunting-Horses" and the multiple-heat races for larger fields of starters.
- The Articles specifying races totalling distances of 8 to 16 miles, plus 12 stone (168 lbs.) to be carried by horses averaging 14:2 hands in height, produced slower races and consequently horses whose average speed was slower.
- To meet the distance conditions of the Articles, breeders turned to various familiar imported mid-Eastern strains. One hundred years earlier, in his *The four chiefest Offices belonging to an Horseman* (c. 1565), Thomas Blundeville had observed: "… those horses that we commonly call Barbarians, do come out of the King of Tunnisland … and able to make a verrie long carrere, which is the cause why we esteeme them so much."
- These sires were small; the Godolphin Arabian, imported 1730, stood 14:1½ hands high (Prior, *Royal Studs*, 1935, p. 135). Line breeding to these imports helped to delay the increasing size of the Thoroughbred racehorse, eventually brought about by better feeding and more careful selection of breeding stock.
- When bred to native mares with sprinting speed bloodlines, many of the Eastern sires contributed high spirits, quality, soundness and stamina, but just two—Place's White Turk and the Godolphin Arabian—provided the middle-distance speed to win King's Plates. Adding this Turcoman racing blood to English sprinting speed produced the new 4-mile winners.

Four-Mile Heat Racing in Nineteenth Century America

Four-mile heat racing in England, given up during the early nineteenth century, continued in the United States until the Civil War (1861). On April 2, 1855, at the New Orleans, Louisiana, Metairie course, Richard Ten Broeck's Lexington 1850, ridden by Gilpatrick, galloped 4 miles in 7 minutes, 19 ¾ seconds, a world's record. America's classic Triple Crown races for 3-year-olds were founded a century later than their British models—the Belmont Stakes in 1867, the Preakness Stakes in 1873, and the Kentucky Derby and Oaks in 1875.

CHAPTER IX

DIPLOMATIC GIFT STALLIONS: BARBS AND TURCOMANS

Morocco Barbs

As noted previously, when bred to Irish and English sprinting mares, the majority of seventeenth century Barb, Turk and Arabian sires contributed high spirits, beauty, quality, soundness, stamina, and the ability to gallop long distances. However, they did *not* contribute the speed required to win multiple heat races over 4-mile courses.

Natural Barbs

The *General Stud Book* lists three types: Barbs bred in England, Natural Barbs and Morocco Barbs. For the word natural, the appropriate equivalent given by the Oxford Dictionary is native. The *General Stud Book* usually uses the word to designate horses native to countries other than England, to mean, in effect, foreign. This word includes stallions known as Barbs imported from Spain, France (via Marseilles) and Italy. An example is the great-granddam of Sir William Ramsden's Byerley Turk Mare (Bruce Lowe #1, *G.S.B.*, p. 5), a Natural Barb Mare of Mr. Tregonwell's, or, more accurately, a mare by Mr. Tregonwell's Natural Barb (J. B. Robertson, *Origin and History of the British Thoroughbred Horse*, 1940, pp. 30–31).

The St. Victor Barb

The only pre-1750 Barb in the *General Stud Book* imported from a country not in North Africa was the St. Victor Barb. The entry in the *General Stud Book* reads (p. 391): "The St. Victor Barb (sire of the Bald Galloway). Brought from France by Captain Rider." Captain Rider lived at Whittlebury Forest in Northamptonshire. The horse took its name from Monsieur St. Victor, presumably the previous owner.

The St. Victor Barb was the only imported Barb, Turk or Arabian, registered in the *General Stud Book*, with enough speed to win races in England. The others were more or less successful at stud, but not on the race course. An item dated March 1682 in the *London Newsletter* reads: "A Match. Mr. Rider's French Horse—1. His Majesty's [Charles II] Corke—2" (Muir, *Newmarket Calendar*, 1892, p. 24). The St. Victor Barb and the Darley Arabian were swept into the mainstream of Thoroughbred bloodlines through matings with mares whose pedigrees were filled with speed strains.

Captain Rider purchased from Lord D'Arcy the Royal mare, Grey Whynot (Bruce Lowe #15, *G.S.B.*, p. 18), by Old Whynot, out of a Royal mare. Old Whynot was by the Fenwick Barb, and also out of a Royal mare. By breeding St. Victor's Barb to Grey Whynot, Captain Rider assured the success and the reputation of his stallion. The foals so produced included the stallion Bald Galloway and Lord Godolphin's mares Points and Cupid. The Bald Galloway appears in the pedigrees of Matchem 1748 and of Eclipse 1764 and also, along with his full sister Points, in the pedigree of imported Janus (1746–1780), the great Virginia sire of quarter-mile racehorses.

Morocco Barbs

The Emperors of Morocco practiced equine diplomacy. Relationships with other countries were cultivated through gifts of beautiful horses to heads of state, of which four stallions are listed in the *General Stud Book*. They are Helmsley's Old Morocco Barb, presented to Charles I in 1637; Dodsworth, presented to Charles II in 1675; Curwen's Bay Barb, presented to Louis XIV of France, later given to his natural sons, Count Byram (Master of the Horse) and Count Thoulouse, and finally acquired by Henry Curwen of Workington Hall in 1698; and the Thoulouse Barb brought to England from France by Henry Curwen and sold to Sir J. Parsons .

Lady Wentworth (*Thoroughbred Racing Stock*, 1938, pp. 235, 236) wrote: "The Moroccan palace of the great Master of the Horse (the Emperor) had three stables for six hundred horses, and one for two hundred of the King's special Arab horses." She cited Don Francisco Alvarez, who had been in "Barbary" from 1520 to 1526, who said:

> Their horses of the countrey breed are in number infinite, but such small hackney jades that they doe them little service, howbeit those that are brought out of Arabia and Egypt are most excellent and beautiful horses.

Lady Wentworth continued:

> Mouvette, a Frenchman who was a slave in Morocco in 1680, tells of Arabian mares of the most beautiful type coming to the Emperor's stables from Egypt and Arabia, for he said:
> "The noble kind which they call 'Asil' are always from thence or of Arabian pedigree and not Barbaries, though they are sometimes wrongly called so if they are bred by the barbarians. These Arab mares and stallions of the 'noble' Arab race are those sought after by the English and French monarchs and are eagerly coveted by the Ambassadors of Europe, for they are indeed of wonderful superiority and price."

As the then most prominent breeder of Arabian horses, Lady Wentworth would not be expected to understate their importance. Because of their great beauty, it is probably safe to assume that the bloodlines of the Old Morocco Barb, Dodsworth, the Thoulouse Barb and Curwen's Bay Barb were, at least in part, Arabian.

The Old Morocco Barb

The earliest of the Barbs exported by the Emperors of Morocco, registered in the *General Stud Book* (p. 14), was the Old Morocco Barb, who stood at the Helmsley stud in Yorkshire. In a letter dated November 9, 1637, the Archbishop of Canterbury noted that an Ambassador from the Emperor of Morocco had just arrived with presents for King Charles I, including four stallions. One of these was probably the Old Morocco Barb. In the July 1649 inventory compiled by Cromwell's Parliamentary committee, Barb strains in the Tutbury stud are represented by five mares and one younger horse whose names contain the word Morocco. Presumably the six 'Morocco' horses in the inventory were descended from one or more of these four stallions.

It is plausible the King gave one of these stallions to the Duchess of Buckingham, who had received Helmsley (1620) as part of her dowry. She had managed the stud since that time, especially after the assassination (1628) of the Duke, who had been Master of the Horse since 1616.

Dodsworth

On the death of James D'Arcy the Elder in 1673, he was succeeded, as Master of the Studs, by Sutton Oglethorpe. It was presumably through Oglethorpe's good offices that the Emperor of Morocco presented a Barb mare to the British Secretary of State, Lord Arlington. Bruce Lowe gave the mare the number 43, the last of his list. At the same time (c. 1675), another mare from Morocco was given by the Emperor to Charles II; she appears under the title *Royal Mare* on p. 15 of the *General Stud Book*. Foaled in 1665 and bred before leaving Tangiers, she is known as the Dam of Dodsworth, the colt foal she produced after arriving at the Hampton Court Royal stud in England. She remained at Hampton Court until she was sold, following the death of Charles II (1685) for 40 Guineas. Her only other recorded foal, by the Helmsley Turk, was the filly Vixen, born 1686 when the mare was 21 years old. Bruce Lowe gives this family the number 32. The *General Stud Book* entry reads (p. 388):

> Dodsworth, though foaled in England, was a natural Barb. His dam, a Barb mare, was imported in the time of Charles the Second, and was called a Royal Mare. She was sold to Mr. Child by the stud-master, after the King's death, for 40 gs., at twenty years old, when in foal (by the Helmsley Turk) with Vixen, dam of the Old Child Mare. [Dodsworth] was the sire of Dicky Pierson (called in some pedigrees The Son of Dodsworth).

Fortunately, Dodsworth did not remain at Hampton Court, but was sent to North Yorkshire. He was bred to good mares in that area, and appears in several late seventeenth century pedigrees. He also appears in the pedigrees of Matchem 1748 and Eclipse 1764 and consequently in the pedigrees of most present-day Thoroughbreds.

Curwen's Bay Barb

The entry in the *General Stud Book* reads (p. 390):

> Curwen's Bay Barb was a present to Lewis the Fourteenth from Muly Ishmael, King of Morocco, and was brought into England by Mr. Curwen, who being in France when Count Byram and Count Thoulouse (two natural sons of Lewis the Fourteenth), were, the former, Master of the Horse, and the latter, an Admiral, he procured of them, about the end of the seventeenth century, two Barb horses, both of which proved excellent stallions, and are well known by the names of the *Curwen Bay Barb* and the *Thoulouse Barb.* Curwen's Bay Barb got Mixbury and Tantivy,

> both very high-formed galloways [ponies], the first of them was only thirteen hands two inches high, and yet there were not more than two horses of his time that could beat him, at light weights;...Brocklesby Betty, and Creeping Molly, extraordinary high-formed mares;...He got two full sisters to Mixbury, one of which bred Partner...and the dam of Crab...He did not cover many mares except Mr. Curwen's and Mr. Pelham's.

Henry Curwen of Workington Hall, Cumberland, was the owner of the Vintner Mare, to whom Bruce Lowe assigned the number 9. In 1688, when King James II abdicated and fled to France, he was followed there by Henry Curwen, also a devout Roman Catholic. We know nothing about the Vintner Mare except that she was black. During Curwen's exile, his horses were sheltered by Charles Pelham, of Brocklesby Park, Lincolnshire. The Brocklesby pack of foxhounds, whose pedigrees are traced to the 1740's, has been famous for 250 years.

According to an inscription on a portrait owned by the Earl of Ancaster (Prior, *Early Records of the Thoroughbred Horse*, 1924, p. 145), Curwen's Bay Barb was brought to Workington Hall in 1698 and died in his 38th year. William Pick (*Turf Register*, p. 15, 1803) wrote:

> The CURWEN BAY BARB was distinguished, for several years, by the bare style [name] of the BAY BARB, and was so well known to Sportsmen by that name, as he would have been, provided there had never been another Barb Horse of his colour in the kingdom.

The Curwen Bay Barb, who stood at Charles Pelham's Brocklesby stud in Lincolnshire, was the most successful stallion of all the imported Barbs. He appears in the pedigrees of Matchem 1748 and King Herod 1758, and is the ancestor of most Thoroughbreds running today.

Accompanying the Curwen Bay Barb and from the same source was a somewhat larger stallion, the Thoulouse Barb (*G.S.B.*, p. 391). Sold to Sir J. Parsons, this horse appears less frequently in early Thoroughbred pedigrees. The importance of these four documented importations in the evolution of the Thoroughbred is indicated by their inclusion in the pedigrees of the three *Great Progenitors* of the tail male lines, Matchem, King Herod and Eclipse.

King William III's Barb Importations

In 1699 Richard Marshall, Stud Master to William III, shipped from Tunis in North Africa (possibly a gift from the Bey of Tunis) nine Barb stallions and five Barb mares which were placed in the Royal Stud at Hampton

Court. During the early eighteenth century, most of the best Thoroughbred brood mares and stallions were in North Yorkshire. The mares at Hampton Court were of lesser quality. In consequence this importation had little effect on the evolution of the breed. The most influential stallions were those presented to other breeders. In 1700 the King gave a grey Barb stallion to John Hutton Sr. of Marske (1659–1736), near Richmond in North Yorkshire, father-in-law of Lord D'Arcy's daughter, Betty D'Arcy Hutton. Her son, John Hutton III, was the breeder of the stallion Marske 1750, sire of Eclipse 1764. Hutton's Grey Barb appears in the pedigree of Eclipse and of other important sires and dams.

From 1725 to 1750 the popularity of Arabian sires was bolstered by the racing exploits of Flying Childers 1715–1722, son of the Darley Arabian imported 1704. The importation of Barb sires was virtually abandoned as horses with crosses of Arabian blood and native English sprinting speed strains became increasingly successful and popular.

Suleiman the Magnificent, Sultan of the Ottoman (Turkish) Empire 1520–1566, by Flemish artist Hans Eworth (fl. 1540–1574), painted 1549. *The Sultan is mounted on a stallion of the type called a Turk or Turcoman, imported from the eastern Mediterranean during the sixteenth and seventeenth centuries by British racehorse breeders.*

Turcoman-Arabian Stallions with Middle Distance Bloodlines Presented by the Sultan of Turkey as Diplomatic Gifts

The ancient district of Turcomania is a high altitude semi-arid grazing area between the Caspian and Black Seas. Since before 1000 B.C., horses of this region had been a racing breed over various middle distances.

The Ottoman Turks captured Aleppo in 1516. The Sultans were so impressed by the beauty of the Arabian horses that their export was strictly forbidden. Bedouin-bred Arabians (Asil) were, and still are, the world's fastest breed of horses over long (50 to 100 mile) distances, speed acquired during centuries of desert warfare. Arabian horses of districts east and north of Aleppo were bred for racing lesser distances for 2,000 years prior to importation to England. No Arabian imported during the seventeenth century is registered in the *General Stud Book*. At age 74, the widely traveled Duke of Newcastle had seen "but one right Arabian" (*A New Method and Extraordinary Invention to Dress Horses,* London, 1667, p. 63). British breeders nevertheless believed that Arabian sires would provide the middle distance speed over multiple heat 4-mile courses needed to win King's Plate races founded 1665 by King Charles II.

A request by Charles II for an Arabian stallion was ignored by Sultan Mohammed IV. He was bound by the prohibition against the export of Arabians. In a letter to Earl Lesley, written at Pera, February 24, 1666, by

Diplomatic Gift Stallions: Barbs and Turcomans

the British Ambassador to Turkey, Lord Winchelsea, said:

> Ever since your Ex. departed home I could not buy one good horse, partly because ye Court is at Adrianople and ye Viziers going into Candia [Crete], for now all ye good horses are either at Court or Morea nigh to ye seaside. I do intend to try if any good ones are to be procured at Byk Be Zar, 13 dayes journey hence, of Turcoman-bredd, for from Aleppo I do despaire of getting any Arrabs as yet, my correspondent there cannot get any for the King. As soon as I can get any yr Exclle. shall have ye best which I will send to ye Resident. (Quoted in *Thoroughbred Racing Stock*, by Lady Wentworth, 1938, p. 265)

Focal point of a horse-racing culture tracing back over two millennia, Turcoman horses had developed bloodlines which contained middle distance speed. Since Aleppo Arabian exports were forbidden, the Grand Signors (Turkish sultans) previously had adopted Lord Winchelsea's alternative. Instead of Arabians as diplomatic gifts, they presented Turcoman stallions, their beauty sometimes enhanced by Arabian bloodline crosses. It was a fortunate circumstance that Turcoman bloodlines provided the middle distance speed required to win King's Plates. This was the third and final source of Thoroughbred speed.

Place's White Turk

Oliver Cromwell, the Lord Protector, was anxious to improve the bloodlines of British bloodstock. He deplored the destruction of the Tutbury Royal stud in 1650 and the dispersal of the mares with Irish Hobby and English Running-Horse sprinting speed bloodline strains. Early in 1657, presumably through his ambassador to Turkey, Sir Robert Bradyshe, Cromwell asked Sultan Mohammed IV for a diplomatic gift of an Arabian stallion from Aleppo. Cromwell and other British breeders apparently were unaware that the export of Aleppo Arabians was strictly prohibited, a prohibition applied to Sultans as well as everyone else.

Mohammed IV replied to Cromwell with the standard alternative diplomatic gift substituted by Sultans during the seventeenth century. This was a Turcoman-Arabian stallion whose pedigree had one or more Arabian crosses. The horse was accepted with diplomatic courtesy. In early November 1657, a light grey stallion, accompanied by Nicholas Baxter, "His Highness' Gentleman of the Horse," was shipped from The Brill, Holland, on the warship Dartmouth, Captain Richard Booth, and landed at Gravesend, England. The new arrival, probably housed in the

Royal stables at Hampton Court stud, was placed in the care of Cromwell's Horse Master, Rowland Place.

Ten months later, on September 3, 1658, Oliver Cromwell died. It was soon evident that the government of the Commonwealth could not long survive the death of its leader. In 1659, the year before the Restoration of Charles II, the Parliamentary government was in shambles. Rowland Place's family estate, Dinsdale, adjacent to the North Yorkshire studs, would be an ideal location for the stallion, but there was no functioning official authority from which Place could ask and receive permission for the move.

Because he and the stallion had served Cromwell, Place was already hated by the Cavaliers, supporters of Charles II. If he took the stallion, to whom he had become deeply attached, to his own home without permission, Place would be hated as a thief by the Roundheads, supporters of Cromwell, and the stallion would thus be branded as stolen goods. Place's personal career would be ruined.

Doubly-damned, Place courageously moved the grey stallion to Dinsdale, adjacent to the town of Richmond and the Vale of Bedale in North Yorkshire. As his grey coat became paler, the stallion was thereafter known as Place's White Turk. The entry in the *General Stud Book* (p. 388) reads simply: "The property of Mr. Place, Stud-Master to Oliver Cromwell when Protector."

In 1657, the Second Duke of Buckingham had regained ownership of the Helmsley stud and its Hobby sprinting speed strain mares. Since 1651, James D'Arcy the Elder had collected and taken to his Sedbury stud Running-Horse and Hobby sprinting speed strain mares from the Welbeck, Wallington and Tutbury studs, whose sprinting speed mares had been dispersed and scattered during the English civil wars 1642–1660. Both Helmsley and Sedbury were in North Yorkshire and so were the rest of the top brood mares in England.

Fortunately Dinsdale was so situated that England's leading breeders of racehorses, with stud farms in the area, could conveniently come to see the handsome grey stallion. A few years after the arrival (c. 1659) of Place's White Turk at Dinsdale, a number of so-called Turks began to appear in North Yorkshire and in south County Durham. These were his descendants.

Place's determination to save the White Turk from obscurity at the cost of his own reputation was completely vindicated. During the next 16 years, he overcame formidable obstacles to make it possible for this stallion to attract the top-class mares whose offspring made him the most influential sire of the seventeenth century.

The Campaign Against Rowland Place

Horsemen stopped using the name of Rowland Place. The D'Arcy White Turk is registered on page 389 of the *G.S.B.* and Place's White Turk on page 388. When writing pedigrees James D'Arcy the Younger called the first, his own horse, "My White Turk," but called the second "The White Turk." Place's name was ostracized.

In 1727 John Cheny compiled the first of the annual *Racing Calendars* listing racing results for the year at the various racecourses. In 1743, presumably to improve sales, he added pedigree information, much of which included the second half of the seventeenth century. In 1791, when James Weatherby compiled *An Introduction to a General Stud Book*, he copied Cheny, word for word. It was not until 1743 that the name of Place's White Turk, imported in 1657, appeared in the *Racing Calendar* (p. xii). The entry was "damned with faint praise." As Place's White Turk's get, Cheny mentioned only two minor sons, Wormwood and Commoner. This listing is repeated in every edition of Volume I of the *General Stud Book*.

Four important late seventeenth century sires with the name 'Turk' are registered in the *General Stud Book*, purposely without pedigrees-the D'Arcy Yellow Turk, My White Turk, the Second Duke of Buckingham's Helmsley Turk and Captain Byerley's Byerley Turk (not imported, Prior, *Early Records of the Thoroughbred Horse*, 1924, p. 143). Printing the names of their sires would have revealed and honored their tail male ancestor, Place's White Turk, and the politically hated Rowland Place of Dinsdale.

Place's White Turk: Middle Distance Speed

Place's White Turk, imported in 1657, was up to that time the sole genetic source of the middle distance speed required to win multiple 4-mile heat course King's Plates in the seventeenth century. He was undoubtedly the sire of the D'Arcy Yellow Turk. Spanker (*G.S.B.*, p. 14) was by the D'Arcy Yellow Turk out of the Old Morocco Mare. Spanker's dam was by the Old Morocco Barb out of Helmsley's Old Bald Peg, prime source of Hobby strain sprinting speed. Spanker's paternal grand-dam, the dam of the D'Arcy Yellow Turk, was a Sedbury mare with a sprinting speed background.

The *G.S.B.* (p. 14) says Spanker "was the best [race] horse at Newmarket in Charles II's reign," the best because he won the most Newmarket King's Plates. Of Spanker's four grandparents, one was a Barb, two were sprinters. Spanker got his middle distance speed from his fourth grandparent, the only source of this speed, Place's White Turk.

The color of Spanker's sire, the D'Arcy Yellow Turk (by Place's White Turk), not characteristic of the European breeds, is seen frequently in Turkmenistan horses, including the Akhal-Teke breed. The color is also called 'golden.'

The D'Arcy Yellow Turk

Evidence of Arabian crosses in the pedigree of Place's White Turk is found in the *General Stud Book*. After the death of the Second Duke of Buckingham, Charles Pelham of Brocklesby Park, Lincolnshire, who became the owner of Spanker, changed the name of this great racehorse and sire to Pelham's Bay Arabian (*G.S.B.*, p. 14).

The introduction of Part IV of *G.S.B.* (p. 388) listing Barb, Turk and Arabian stallions, notes that "some of them were not imported at all, it being common [practice] to call horses Barbs, Turks, [Arabians] etc., that were only of that breed," meaning only that the stallion had an imported ancestor. Spanker was the son of the D'Arcy Yellow Turk whose imported sire, Place's White Turk 1657, had an Arabian cross in his pedigree. When Charles Pelham changed Spanker's name to the Pelham Bay Arabian, he was following the common practice of his time, attributing the Arabian designation to any handsome horse with an eastern provenance, regardless of origin.

The Old Morocco Mare produced four foals, a colt, followed by three fillies. The D'Arcy Yellow Turk was the sire of the colt (Spanker), foal number 3 (filly) and grandsire of number 4 (the Spanker Mare, incestuously, by Spanker). Foal number 2, Young Bald Peg was by Leedes Arabian, bred by Edward Leedes of North Milford, who appears frequently in early pedigrees, in all probability, a son of the D'Arcy Yellow Turk. Leedes acquired foals number 2 and 4, who became his foundation mares.

After the death of James D'Arcy the Elder (1673), his post as Master of the Studs to Charles II was filled by Sutton Oglethorpe. The latter was the owner of a stallion by the D'Arcy Yellow Turk. Once again, following the practice of the time, the horse was called the Oglethorpe Arabian (*G.S.B.*, p. 389).

In her *Thoroughbred Racing Stock* (1938, p. 389), Lady Wentworth said the D'Arcy Yellow Turk was "by far the greatest male influence in Thoroughbred pedigrees" and the D'Arcy White Turk was next in influence. Lady Wentworth mistakenly believed these two stallions to be Arabian.

Midridge Grange, property of Captain Robert Byerley in South County Durham, where the Byerley Turk stood at stud.

The Byerley Turk

The fourth major British-bred seventeenth century sire presumably descended from Place's White Turk was the Byerley Turk, the property of Captain Robert Byerley (1660–1714) of Midridge Grange, south County Durham, not far from the stable of Place's White Turk at Dinsdale. The Byerley Turk was not an imported horse.

In his *Early Records of the Thoroughbred Horse* (1924, p. 143), C. M. Prior cites the records of Captain Byerley's military career and as a member of Parliament, which explode the conjecture that his Turk was a prize of war, captured from the Turks by Byerley at the Siege of Buda (Hungary) in 1687. There is no record that Captain Byerley took part in the Siege of Buda. All the *G.S.B.* tells us (p. 389) is that he was "Captain Byerly's [sic] charger in Ireland in King William's wars (1689, etc.). He did not cover many bred mares [pedigreed mares]."

In 1690 after resigning his commission as commanding officer of the Dragoon Guards, Col. Byerley took his charger back to Midridge Grange. Here the horse began his successful career at stud in 1691. He was still active in 1702 when he sired Sir William Ramsden's bay colt Basto, out of Edward Leedes' Bay Peg. After an excellent racing career, Basto was premier sire at the Duke of Devonshire's Chatsworth stud. The colloquial name for Basto was "Byerley's Treasure."

The Byerley Turk,
a Fores engraving after
John Wootton (1683–1764).
This portrait of the Byerley Turk was probably not from life. The Byerley Turk was not an imported horse (C.M. Prior, Early Records, 1924, p. 143). He stood at stud from 1691 to c.1702.

The Byerley Turk is registered as the earliest of the three tail male Thoroughbred bloodlines by direct descent through King Herod 1758 to The Tetrarch 1911, sire of the famous matron Mumtaz Mahal 1921. He also appears in the pedigrees of the other two Great Progenitors, Matchem 1748 and Eclipse 1764. Further details of Place's White Turk and the Byerley Turk appear in Chapter XI in connection with King Herod 1758, bred by the Duke of Cumberland.

Lister and Holderness Turks

Three other imported horses in this period were called Turks and registered in the *General Stud Book*: the Lister Turk 1687, the Holderness Turk 1704 and the Belgrade Turk 1717. The last is considered in Chapter X, "The Godolphin Arabian."

The Lister Turk

The Lister Turk is the only Turk prize of war entered in the *General Stud Book* who appears in the pedigrees of all three Great Progenitors: Matchem, King Herod and Eclipse. The *G.S.B.* entry (p. 389) reads: "The STRADLING or LISTER TURK was brought into England by the Duke of Berwick, from the siege of Buda [Hungary (1687)] in the reign of James the Second." This grey stallion, first known as the Stradling Turk, became the property of Matthew Lister of Burwell Park, near Louth, in Lincolnshire which is next to North Yorkshire.

The Lister Turk appears five times in the pedigree of Eclipse, three times as the sire of Lister's Snake and twice as the sire of the Duke of Rutland's Coneyskins (grey) foaled 1712. The dam of Lister's Snake was by D'Arcy's Hautboy (by D'Arcy White Turk) and the granddam of Coneyskins was by Spanker (D'Arcy Yellow Turk ex Old Morocco Mare). It was the speed bloodlines of these two mares that helped to move the Lister Turk into the mainstream of Thoroughbred bloodlines.

The Holderness Turk

An entry in the *General Stud Book* (p. 390) reads: "HOLDERNESS TURK (sire of Hartley's Blind Horse), was brought from Constantinople by Queen Anne's Ambassador about 1704." The stallion was acquired by the Earl of Holderness, a relation of the D'Arcy family. No attempt to represent the horse as an Arabian seems to have been made. Had it been a diplomatic gift from Sultan Ahmed III, a Turcoman-Arabian stallion, this would have been mentioned in the *Stud Book*. In such case the horse would probably have remained at Hampton Court stud. Bred to D'Arcy mares, the Holderness Turk achieved some success, but did not influence the evolution of the Thoroughbred breed.

CHAPTER X

THE GODOLPHIN ARABIAN 1724-1753

"This Extraordinary Foreign Horse"

The year 1731 was made memorable by the death of James D'Arcy the Younger of Sedbury, born in 1650 and created Lord D'Arcy of Navan in 1721. This brought to an end the Sedbury stud in North Yorkshire which, for the previous 80 years, had dominated British bloodstock breeding. Sedbury was the last of the great stud farms whose bloodlines were based primarily on sprinting speed strains. In the same year another event, then little noticed, proved to be even more significant. The Godolphin Arabian went to stud. It was this great sire who contributed the final strains to the bloodlines of the Thoroughbred breed, influencing both conformation and the ability to maintain speed "over a distance of ground." He was by far the most influential of all the sires who flourished during the evolution of the Thoroughbred racehorse before 1750. The extent to which he remolded the breed was miraculous.

Published Appraisals 1753–1808

The Godolphin Arabian's career as a sire (1731–1753) was festooned with sons and daughters who were top-class racehorses and who won additional fame as stallions and brood mares. It is surprising that no general appraisals of this illustrious career were published until some 40 years after his death. He died on Christmas Day, 1753, and is buried under an archway leading to the stable courtyard of Gog Magog Hills, the stud farm of the Second Earl of Godolphin.

The Godolphin Arabian, an engraving by John Scott published in 1809 after George Stubbs (1724–1806).

Stubbs' painting was based on a 1753 engraving by John Faber (1685?–1756) after David Morier's (1704–1770) replica of his portrait of the Godolphin "Taken from life." Morier's first painting was commissioned by H.R.H. the Duke of Cumberland, breeder of Herod 1758 and Eclipse 1764. The Godolphin was not an Arabian; he came from a strain of horses used by the Turkish sultans for diplomatic gifts. These gift horses were Turcoman stallions with Arabian bloodlines added for enhanced beauty. The print shows typical Turcoman deep, sloping shoulders, short back, heavily muscled hindquarters, as described in Osmer's Dissertation on Horses. (An earlier example of this strain is Place's White Turk, G.S.B., p.388, presented to Oliver Cromwell by Sultan Mohammed IV in 1657.) Since the horse had Arabian crosses in his pedigree, and in accordance with eighteenth century common practice (G.S.B., p.388), his importer, Edward Coke (1730) called the Godolphin an Arabian. This enabled Coke to charge a much higher stud fee.

In 1751, Reginald Heber had taken over the *Racing Calendar* from its 1727 founder, John Cheny. One would expect to find in the *Racing Calendar*, published in 1754 and 1755, an obituary of this great stallion. Its omission is explained by the print published at the time by Heber which was advertised for sale in the 1754 and 1755 Calendars. This was a 1753 engraving by John Faber after the portrait of the Godolphin Arabian painted from life by David Morier. At the bottom of the print is a summary of the Godolphin's career at stud. It is likely that the publication of an obituary in the *Racing Calendar* would have hurt the sale of the print.

Weatherby, Pick, Taplin: 1791–1808

C. M. Prior called William Pick "the most diligent and painstaking of all turf historians" (*Royal Studs*, 1935, p. 127). In 1786 Pick published his first *Turf Register—An Historical Account of the Most Favourite Arabians, Turks and Barbs*. Strangely, the Godolphin Arabian was not included. It was not until 38 years after his death that the extraordinarily successful stud career of the Godolphin Arabian was outlined in a racing publication. The compiler was James Weatherby, since 1773 proprietor of the *Racing Calendar* and since 1774 Keeper of the Match Book at Newmarket. In 1791 Weatherby tried an experiment. He published a slim volume entitled *An Introduction to a General Stud Book*. It was spectacularly successful, the foundation pedigree volume of the Thoroughbred breed. The first of these volumes is now more than 200 years old. It is reputed to be the first published stud book of any breed of livestock.

The entry of the Godolphin Arabian (*G.S.B.*, 1791, pp. 206–207) is as follows:

> GODOLPHIN ARABIAN. Of this valuable stallion (strange as it would undoubtedly appear) scarce any records are extant; all that can be discovered, after strict enquiry, is, that he was a brown horse, about fifteen hands high, that he was first the property of Mr. Coke, and given by him to Mr. Roger Williams, keeper of the St. James Coffee-House, by whom he was presented to Ld Godolphin, and that he continued in his Lordship's possession, as a private stallion, till his death. To those who are thoroughly conversant with the Turf, it would be superfluous to remark, that he undoubtedly contributed more to the improvement of the breed of horses in this country, than any stallion before or since his time: it would be equally unnecessary to enumerate his get: to those who are less acquainted with the annals of racing, the names of Cade, Regulus, Blank, Babraham, and Bajazet, may serve as a

> proof of the remark: and it may not be amiss to observe, that almost (if not entirely so) every SUPERIOR horse of the present day, partakes of his valuable blood.—He died at Hogmagog [sic], in 1753, in the 29th year of his age, and is buried in a covered passage, leading to the stable, with a flat stone over him, without any inscription.—In regard to his pedigree, from all that can be collected, none was brought over with him, as it was said, and generally believed that he was stolen.—It may appear trifling to notice the extraordinary affection shown by this horse to a cat, who lived in his stable, which was more particularly manifested by his extreme inquietude on the death of that animal. We mention this circumstance merely to account for the introduction of a cat in the portrait of the Godolphin Arabian, to which the reader is referred for an accurate representation of him.

Weatherby's appraisal of the Godolphin is contained in the sentence: "He undoubtedly contributed more to the breed of [race] horses than any stallion before or since his time—almost (if not entirely so) every SUPERIOR horse of the present day, partakes of his valuable blood."

Two publications of 1803 repeated Weatherby's 1791 entry in the *Introduction to a General Stud Book*. These were William Pick's invaluable *Turf Register* (Vol. I, p. 57) and William Taplin's *Sporting Dictionary* (Vol. I, p. 51). The latter wrote of the Godolphin, "there can be no doubt, from the success of [his] progeny…but that he contributed more to the value and speed of horses for the turf, than any other foreign stallion ever brought into this kingdom."

Subsequent printings of Weatherby's *An Introduction to a General Stud Book* appeared in 1793, 1800, 1803 and 1808, demonstrating its general popularity and its great contribution to breeders and to the art of breeding racehorses. The edition of 1808, which included important additions from Pick's *Turf Register,* is considered the first more-or-less complete edition of Volume I. Weatherby's remarks (p.516) include in part:

> The GODOLPHIN ARABIAN…there is a picture of him and his favourite Cat, in the library at Gog Magog in Cambridgeshire…That he was a genuine Arabian, his excellence as a Stallion is deemed as sufficient proof. In 1731, then the property [of] Mr. Coke, he was Teazer to Hobgoblin, who refusing to cover Roxana, she was put to the Arabian, and from that cover produced Lath, the first of his get. It is remarkable that there is not a superior horse now on the Turf, without a cross of the Godolphin Arabian, neither has there been for several years past.

> Mr. Coke is said to have imported the above Arabian from France, and the Editor was once informed by a French gentleman, whom he has not had an opportunity of seeing since, that this horse had actually drawn a cart in the streets of Paris.

Hobgoblin did not go to stud until 1735, two years after Coke's death. It is quite possible the story was true but the reluctant stallion was actually another horse. Weatherby must never have found the Frenchman who could corroborate the cart story. He did, however, repeat the statement "that there is not a superior horse on the turf without a cross of the Godolphin Arabian," to which he added, "neither has there been for several years past."

The Godolphin Arabian's Career at Stud

Manuscript Stud Book

In Volume I of the *General Stud Book*, James Weatherby lamented his inability to locate the manuscript (hand-written) records of individual breeders which would have been included in his compilation. C. M. Prior, 133 years later, did better. In 1924, he published *Early Records of the Thoroughbred Horse* and in 1935 *The Royal Studs of the Sixteenth and Seventeenth Centuries*. The culmination of Prior's extensive and valuable additions to early Thoroughbred history came to light in the late 1920s when the Duke of Leeds, direct descendant of the Second Earl of Godolphin, ordered the removal of the contents from Hornby Castle in North Yorkshire. Fortunately a local turf historian, Capt. R. H. Radford, recognized the importance of the manuscript stud book of the Second Earl of Godolphin (1678–1766). C. M. Prior was privileged to publish the entire book with explanatory notes on pages 127 to 178 of his *Royal Studs*. The reproduction of the manuscript stud book begins on page 137. The end of the manuscript's stud book reads:

> The Godolphin Arabian was a brown bay, the off hind foot white, and a little white on ye inside corner of ye near Hind Hoof. He dyed ye 25th off Decr, 1753. He was 14 Hands, one Inch & half, without shoes.

Longford Hall, 1730–1733

The Godolphin manuscript stud book includes the years 1732 to 1756. Although the Godolphin Arabian changed owners during this period, it is all in the same beautiful handwriting, a model of penmanship.

The 1808 edition of Vol. I of the *General Stud Book* reads: "Mr. Coke is said to have imported the Arabian from France." This was Edward Coke (1701–1733), a younger brother of Thomas, known as "Coke of Norfolk" and created Earl of Leicester in 1744. In 1727 Edward had inherited Longford Hall, a fine estate in Derbyshire. The activities of his racing stable from 1727 to 1733 are recorded in John Cheny's *Racing Calendars*. The Godolphin manuscript book shows that in 1730 Coke had six brood mares, the sire Whitefoot and the future sire Hobgoblin.

Coke's best mare was Roxana 1718. Her pedigree contained two crosses of Sedbury Royal Mares, going back in tail female to a mare by Spanker. She won three races for Coke in 1727, the last year she raced. Coke bred her to the Godolphin Arabian in 1731 and 1733, and in 1732 to the Duke of Devonshire's private stallion, the sensational racehorse Flying (Devonshire) Childers. The 1731 mating produced the Godolphin's first foal, Lath. In April 1737 at Newmarket Lath won the 1000 Guineas Great Stakes and later races. Pick (*Turf Register*, 1803, p. 53) said "Lath was allowed to be one of the best horses that appeared at Newmarket since the time of Childers," more than 15 years earlier. The produce of the 1733 mating was Cade, so-called because he was raised on cow's milk, and bottle-raised foals were referred to as 'cades.' He was a very popular stallion in Yorkshire, and the sire of Matchem 1748, the progenitor who carried on the tail male line of the Godolphin. The Flying Childers colt named Roundhead, foaled 1733, was the sire of Joseph Andrews who covered a host of Thoroughbred and Roadster mares in North Yorkshire and adjacent Lincolnshire.

Edward Coke died in August 1733, when he was only 32 years old. Unfortunately he did not live to admire the success of the three matings he arranged during his last 2 years. His will, dated December 11, 1732, contained the following clause: "I give to the Rt. Hon. Earl of Godolphin all my running horses and running mares and studd mares, and to Mr. Roger Williams all my stallions." The stallions were Whitefoot, Hobgoblin and the stallion who became known as the Godolphin Arabian. It was probably to help Williams financially that Coke left the stallions to him. Williams evidently promptly transferred them to Lord Godolphin. By autumn of 1733 all of Edward Coke's horses were moved from Longford Hall to Lord Godolphin's Gog Magog Hills stud in Cambridgeshire, not far from Newmarket, the racing center of England.

The Gog Magog Hills Stud 1733–1753

The clause in Edward Coke's will was a landmark in the evolution of the Thoroughbred. Written by a 32 year old sportsman who had been active in racing and breeding only since he was 25, Coke ensured his bloodstock would not be dispersed. By leaving all his stallions to Williams and brood mares to the Earl, he kept his bloodstock together. When Williams transferred his group to Godolphin, in effect he placed the name *Godolphin* in front of *Arabian*. It also established the Second Earl of Godolphin as one of the great breeders of Thoroughbred history. Lath and Cade, and their successes on the race courses and at stud, established the importance of the Godolphin Arabian as a sire at the very beginning of his stud career.

Obstacles Overcome

The Godolphin Arabian's career at stud faced several handicaps. He was maintained as a private stallion, and only a few friends were invited to breed to the great stallion. He was located in an area that was not populated by the best racing mares, an area far from the center of racehorse breeding in North Yorkshire.

The most eminent of these few outside mares bred by the Godolphin Arabian was Lord Chedworth's Grey Robinson. In 1739, she produced a colt and in 1743 a filly. The colt was the unbeaten Regulus, and the filly was the tail female ancestress of St. Simon. Both trace back in tail female to Lord D'Arcy's Grey Royal (Bruce Lowe #11, *G.S.B.*, p. 10). Other privileged mare owners were the Duke of Devonshire and Mr. Panton. Otherwise the Arabian was bred exclusively to his Lordship's mares at Gog Magog Hills.

Before the Godolphin went to stud, virtually all the stallions who became successful sires stood in the Richmond-Vale of Bedale areas, embracing North Yorkshire and adjacent parts of Lincolnshire and County Durham. Here were the stud farms with the best mares. It is a matter of record that only when bred to such mares could most stallions achieve success as a sire. But Gog Magog Hills was far from Yorkshire. The success of the Godolphin Arabian as a sire was not dependent on geography. He was one of the very few stallions in history so prepotent that success was achieved with mares who were less than the best. In his 1756 *Dissertation on Horses*, William Osmer, a London veterinary surgeon, wrote the classic conformation description of the Godolphin Arabian. (See "Conformation of the Godolphin Arabian," this chapter.) He complained: "It was a pity he was not used more universally on better mares" (cited by Roger Longrigg, *The History of Horse Racing*, 1972, p.61, note 16). The inscription on the dated Morier portrait at Houghton Hall, praises the Godolphin for "hitting it off with most of our mares." A detailed examination of the entries in the manuscript stud book confirms the conclusion that the Gog Magog mares were only of average quality.

Entries in the manuscript stud book indicate that the Gog Magog brood mare band seldom exceeded six or seven mares. It also records that during his 22 years at stud about 80 foals by the Godolphin Arabian were born at Gog Magog. The noted turf scholar, J. B. Robertson ("Mankato") discovered about ten more in the pages of the *General Stud Book*. His annual crop was thus a very modest four or five foals a year.

Throughout Thoroughbred history, certain breeders have repeatedly tried to establish the reputation of a sire by breeding him only to half a dozen select mares, stakes winners and/or producers. Most of these experiments have failed. For some genetic reason, to achieve success, sires have needed mares in quantity as well as quality. The Godolphin Arabian needed neither. He overcame with ease a series of handicaps, any one of which would have ruined the careers of lesser stallions, notably average quality mares and a location far from the best breeding farms.

The Morier Godolphin Arabian Portraits

The Cumberland Lodge Portrait
In 1743 H.R.H. the Duke of Cumberland brought David Morier (1704–1770) from Switzerland to paint portraits of racehorses for his picture gallery in Cumberland Lodge, Windsor Great Park. Morier thereafter signed himself "D. Morier, Painter to H.R.H. the Duke of Cumberland." Morier's portrait from life of the Godolphin Arabian was to hang at Cumberland Lodge for the Duke, not for Lord Godolphin, who had a 1731 portrait of this stallion by John Wootton (1683–1764). Unfortunately, if the Cumberland Lodge picture is still in existence, we do not know its present location.

The Houghton Hall Portrait
Shortly after the Godolphin's death on Christmas Day 1753, the Duke of Cumberland apparently instructed Morier to paint a small replica of the Cumberland Lodge portrait. This 17-inch x 20-inch picture, now hanging at Houghton Hall, was for the publisher Reginald Heber, since 1751

proprietor of the *Racing Calendar*. From this John Faber (1685?–1756) engraved a "large size" fine print (14 ½ inches x 14 inches) for which there was great demand. A line below the picture notes:

"*D. M. pinxt.*   *I.F. fecit 1753*"

The initials represent David Morier and John Faber. There follows an inscription of 111 words describing the Godolphin Arabian and listing the names of 18 sons and daughters.

This artist's replica painting, commissioned by the Duke, is the most famous and most photographed portrait of the Godolphin Arabian. Located in Norfolk, Houghton Hall is one of the most beautiful early country mansions in England, the property of Lord Chomondeley. The surviving evidence identifies this as the portrait upon which the 1753 print was based. At 17¼ inches x 20 inches, the canvas is considered small. This replica painting by Morier is neither signed by the artist nor dated. For Heber's purpose, they were not necessary. The artist, the engraver and the date are part of the inscription on the print.

The Houghton Hall painting shows an attractive, well-bred, compact and small plain brown-bay stallion, with his protesting ears laid back, standing in a courtyard, facing right. He looks at an open stable door. Across the back of the picture is a high brick wall. On the ground just outside the stable door is the cat Grimalkin, the Godolphin's constant companion. When John Faber copied the painting for the 1753 engraving, he engraved a better looking cat. All critics agree, however, that George Stubbs painted the most handsome cat in his portrait based on the 1753 print. Whatever Grimalkin's looks, the Godolphin was inconsolable when the cat died.

Although his withers are not prominent, the horse is remarkable for his deep and sloping shoulders, short back and powerful loins, fully complying with Osmer's 1756 description. The horse does not resemble a purebred Arabian. This was also the opinion of the Duke of Cumberland, who commissioned it.

The Houghton Hall canvas is surrounded on all four sides by a 135 word inscription. The 1753 print inscription listed offspring. So does the Houghton Hall inscription. It also lists a "new" category, an addition with the names of twelve grandchildren, "the best horses of our present time." All were foaled between 1747 and 1749. They would have completed their racing careers and have established their reputations about 1758 or 1759. These are the dates of "our present time." They also represent the dates when this inscription was written.

The evidence indicates that about 1759 Reginald Heber, with the approval of the Duke of Cumberland, was planning a second edition of the Godolphin Arabian print. The Houghton Hall canvas and a new surrounding inscription were to be copied in an engraving from which the print would be made. A horseman would not have commissioned the Houghton Hall format as a picture to be hung over the mantelpiece. We are looking at a painting done in the format of a print. The replica's "new" inscription in highly visible gold lettering on a dim background corrected the hard-to-read faint script of the 1753 inscription, which straddles the Godolphin coat of arms. The addition to the inscription would make the print "new and up-to-date," appealing to the racing public and to the owners of the 1753 print as well. A further appeal, the last sentence of the inscription read: "This is the original picture" (meaning the artist's own copy of his first painting).

Reginald Heber apparently abandoned the project. Prints based on the second edition of the Houghton Hall picture have not surfaced. Another publisher anticipated Heber. In his essay on the Godolphin Arabian portraits, Fairfax Harrison noted that about 1760 a portrait of the Godolphin was published "as Plate 40 in *The Sportsman's Pocket Companion*, the collection of portraitures of horses engraved by Henry Roberts after drawings by James Roberts (1728–1799) … in London by R. & R. Baldwyn" (*The Roanoke Stud*, Richmond, Virginia, 1930, p. 219). Reginald Heber apparently decided not to compete. Since there were then no copyright laws, Roberts had copied the 1753 print, as have most other artists painting Godolphin Arabian portraits, including George Stubbs (1724–1806) and Francis Sartorius (1734–1804).

All was not lost, however. The later Houghton Hall portrait inscription in combination with the 1753 print inscription, in very few words, present the most penetrating and the most inclusive of all the appraisals of the Godolphin Arabian's achievements as a sire.

Inscription Texts from the Print and the Portrait
Here is the text of the 1753 print inscription:

> The Portraiture of the Bay Arabian, the property of the Right Hon'ble the Earl of Godolphin &c.
> D. M. pinxt.   I. F. fecit 1753.
>
> This extraordinary foreign Horse became a private Stallion as soon as he was landed & got a greater Number of Fine Horses of Just Temper with superior Speed than any Arab did. He was the Sire of Lath, Dismal, Cade, Bajazet, Babraham, Phoenix, Dormouse,

Regulus, Skewball, Sultan, Blank, Slugg, Noble, Tarquin, Blossom ye Godolphin Gelding, Shepherdess, Amelia, and several other Runners, besides Stallions and Brood mares in the Greatest Esteem in England, Scotland and Ireland, &c., &c., &c.

He made his Exit at Hogmagog Hills, December 1753 in the 29 year of his age.

The Houghton Hall Portrait Inscription

From the Morier painting, commissioned by the Duke of Cumberland, the inscription reads:

> The Godolphin Arabian. Esteem'd one of the best Foreign Horses ever brought into England. Appearing so, both from the Country he came from, and from the Performances of his Posterity, they being Excellent, both as Racers and Stallions, and Hitting with most other Pedigrees, and mending ye Imperfections of their shape. He was the Sire, Amongst others, of ye following Cattle, Blank, Bajazet, Babraham, Slugg, Dormouse, Regulus, Cade, Lath, Whitenose, Dismal. And ye Grandsire of ye best Horses of the present Time, as Bandy, Mercury, Dutchess, Amelia, Lord Onslow's Victorious, and Cato, Sir C. Sidely's Martin, Lord Eglintown's Whitefoot, Lord March's Danby Cade, Mr. Fenwick's Byewell Tom, Matchem, and Trajan. *And is allowed to have refreshed English Blood more than any other Foreign Horse ever yet imported.* He did [sic] at Hogmagog in December 1754 [1754 was in error for 1753].
>
> The Original Picture taken at the [Gog Magog] Hills by D. Morier, Painter to H.R.H. the Duke of Cumberland. [An 'original' is today called a replica.]

The two inscriptions, in a very few words, outline the qualities which set the Godolphin Arabian apart from other stallions, which explain his extraordinary success as a sire, and his contributions to Thoroughbred bloodlines. One stallion alone altered and improved the breed.

The key sentence of the 1753 print reads: "This extraordinary foreign horse...got a greater Number of Fine Horses...of Just Temper with superior Speed than any Arab did." The words "just temper" (quiet temperament) were written by a racing man well aware that a nervous, excitable horse often "runs his race" (squanders his energies) before reaching the start. It also compares the get of the Godolphin favorably as compared to the get of the high-spirited Arabians. That the Godolphin "got a greater number of fine horses...with superior speed than any Arab did" was an assertion that the offspring of Arabian sires were far slower than the superior speed of the Godolphin's get. This was in accord with the Arabian cross-breeding experiments, masterminded by the Duke of Cumberland, which proved that line breeding to Arabian bloodlines was producing slower racehorses (Taplin, *Sporting Dictionary*, 1803, Vol. II, p. 208).

Most importantly, Cumberland's sentence asserts, flatly, that the Godolphin was an "extraordinary foreign horse," but not an Arabian. Because he was a private stallion, only a few horsemen had the opportunity to see the stallion and to notice he did not, in any respect, resemble either an Arabian or a Barb. Perhaps for fear of displeasing the Earl of Godolphin, they did not say so publicly. Yet the inscription tells us a great deal:

- The inscription praises the Godolphin "*for performance of his posterity, they being excellent both as racers and stallions.*" The names of twelve grandchildren are also listed. The word *posterity* is used to mean descendants.
- The inscription foretells the extraordinary prepotency of the Godolphin which pervaded the entire breed. In the 1808 edition of the *General Stud Book* (p. 516), James Weatherby noted "*there is not now a superior horse on the turf without a cross of the Godolphin Arabian, neither has there been for several years past.*"
- The inscription continues: "*Hitting with most other pedigrees.*" This refers to the pedigrees of the mares to which he was bred. The only superior racehorses and sires got by the Darley Arabian were Flying Childers 1715 and Bartlett's (Bleeding) Childers 1716, out of Betty Leedes, whose pedigree contained three Hobby crosses (Old Bald Peg, *G.S.B.*, p. 14). The Godolphin sired superior horses out of "*all sorts and conditions*" of mares.
- The inscription continues: "*And mending ye imperfections of their shape.*" A single stallion, the Godolphin Arabian, permanently changed the conformation ('shape') of the Thoroughbred. He passed on to his descendants and to the entire breed its most striking feature, his deep and sloping shoulders. These enabled greater extension at the gallop, and therefore a third and different source of speed "*over a distance of ground,*" then, the King's Plate 4-mile heats distance, and today, the Epsom Derby and Belmont Stakes at 1½ miles distance.
- The inscription concludes: "*refreshed English blood more than any other foreign horse ever yet imported.*" Because he was a complete outcross for every mare to which he was bred, the Godolphin Arabian was a fully effective refresher of the English blood.

The inscription presents the conclusions of a breeder with a profound knowledge of seventeenth and eighteenth century pedigrees, bloodlines and conformation. The Duke of Cumberland was primarily responsible for the key sentences in the two inscriptions which identify the achievements at stud of the Godolphin Arabian. This is supported by Cumberland's involvement in both of Heber's print projects. It was Cumberland's remarkable insight and wealth of information on which these conclusions were based.

Conformation of the Godolphin Arabian

Osmer's 1756 Description

The best contemporary word description of the Godolphin Arabian was written by William Osmer, a veterinary surgeon with an office on Blenheim Street, off Bond Street in London, who published several books on veterinary medicine. Osmer was convinced British breeders of racehorses were paying too much attention to pedigree and too little to conformation. He was fascinated with the conformation of this strange-looking horse who was becoming a highly successful sire. In his first book, *Dissertation on Horses*, published in 1756, Osmer wrote (p. 273):

> According to these principles of length and power, there never was a horse (at least that I have seen) so well entitled to get racers as the Godolphin Arabian: for, whoever has seen this horse must remember, that his shoulders were deeper and lay further into his back, than any horse ever yet seen. Behind the shoulders, there was but a very small space ere the muscles of his loins rose exceeding high, broad and expanded, which were inserted into his quarters with greater strength and power than in any horse, I believe, ever yet seen of his dimensions, viz. Fifteen hands high. (Lord Godolphin's manuscript stud book records the stallion's height as 14 hands, 1 ½ inches.)

The angle of the Godolphin Arabian's sloping shoulders, described by Osmer, distinguishes the Thoroughbred from other breeds. The newly acquired sloping shoulders achieved greater extension at the gallop, longer strides, and therefore faster middle distance speeds.

> There is a certain length...absolutely necessary to velocity [speed]. I say it consisted in the depth and declivity [slope] of the shoulders...if the shoulder be upright [straight] the horse will

not be able to put his toes far before him, but if the shoulders have a declivity in them he can put his toes farther before him and greater purchase of the ground will be obtained at every stride. (Osmer, *Treatise on Diseases and Lameness of Horses*, 1761, 3rd Edition, 1830)

The prepotency of the Godolphin Arabian was truly extraordinary. He passed on, not only to his children and grandchildren, but also to the entire breed, the deep and sloping shoulders, the short back and powerful loins described by Osmer.

The chest cavity which accompanied the deep and sloping shoulders was considerably larger than the chest cavities accompanying the straight shoulders of the sprinting speed strains derived from sixteenth and early seventeenth century Hobby and Running-Horse strains. These strains appear in the pedigrees of sprinters whose straight shoulders are illustrated in many eighteenth century portraits. There is, for example, George Stubbs' portrait of Marske 1750 (sire of Eclipse 1764). This shows a stallion of typical sprinter type whose pedigree is dominated by Helmsley stud Hobby strains and by Sedbury stud Running-Horse strains. Greater chest cavity is accompanied by larger lungs whose intake of oxygen is both quicker and superior. The Godolphin Arabian contributed to the Thoroughbred an intake of oxygen capacity unmatched by any other breed of horse.

Prominent Withers

Thoroughbred portrait painters of the eighteenth and early nineteenth centuries record another conformation feature shared by descendants of this remarkable sire. It was also the Godolphin Arabian who contributed prominent withers to the Thoroughbred breed. Portraits of horses with one cross of the Godolphin, such as Eclipse 1764, show good but not prominent withers. Many eighteenth century portraits of horses with two or more crosses, however, show big slashing shoulders topped by the prominent sharp withers characteristic of today's entries in the Liverpool Grand National Steeplechase. In the Paul Mellon Collection, Upperville, Virginia, further pictorial evidence supporting this hypothesis is seen in William Shaw's portrait of King's Son of Blank, foaled in 1764. A present-day owner might point this horse either toward the Liverpool Grand National Steeplechase or toward conformation classes in major horse shows. He is a magnificent animal, full of quality. King's Son of Blank was by Blank, his dam by Regulus, the rest of the pedigree is

King's Son of Blank 1764 by William Shaw, oil.
Both his sire and dam were by the Godolphin Arabian. The horse's remarkable quality and high-set tail reflect this intense inbreeding and the Arabian crosses in the bloodlines of the Godolphin Turcoman-Arabian.

D'Arcy Sedbury sprinting stock, not noted for good looks. There are no Darley or other Arabian crosses. He is closely line bred; both Blank and Regulus are by the Godolphin Arabian. The only plausible source from which King's Son of Blank could have derived his striking quality was the Godolphin Arabian. Another example is the famous Stubbs' portrait of *Hambletonian Being Rubbed Off* (1792).

High Hindquarters

In his word portrait of the Godolphin, William Osmer wrote:

> Behind the shoulders, there was but a small space ere the muscles of his loins rose exceeding high, broad and expanded, which were inserted into his quarters with greater strength and power than in any horse, I believe, ever yet seen of his dimensions. (14:1 ½ hands)

This conformation feature has been observed in certain of his descendants ever since. It appears in George Stubbs' preliminary sketch of Eclipse 1764 and in Edward Troye's meticulously accurate portraits of Lexington 1850. Since the advent of photography, examples are more numerous: St. Simon 1881, The Tetrarch 1911, Hyperion 1930, Nasrullah

1940, Secretariat 1970, and many others. The conformation of these high hindquarters graphically illustrates the present-day continued influence of seventeenth and eighteenth century bloodlines.

When Charles II in 1665 established the 4-mile multiple heat standard, the breeders of short course Hobby and Running-Horse sprinting strains had to look for an outcross which the native Hunting-Horse strains could not provide. They turned to imported mid-eastern sires: Barbs in the seventeenth century, Arabians in the eighteenth century. However, these were long distance gallopers, not fast racehorses. Breeding Barb and Arabian sires to short speed mares produced a few spectacular racehorses, such as Spanker c. 1670, and Flying Childers 1715. The average results, however, were poor. This cross was too extreme. The genes of the Godolphin Arabian, on the other hand, with his unique conformation, his sloping shoulders and his larger lungs, were calibrated to get 4-mile multiple heat racehorses.

The Origin and Importation of the Godolphin Arabian

Who sold the Godolphin Arabian (1724–1753) to Edward Coke? Only a few of the author's answers to this question are based on hard evidence. The rest of the answers are supported only by conjecture—plausible conjecture, based on circumstantial evidence.

In his 1935 book, *The Royal Studs*, C. M. Prior tackled the problem (p. 131). Edward Coke's sister, Cary, was the wife of a prominent racing man, Sir Marmaduke Wyvill of Constable Burton, North Yorkshire, whose ancestral namesake's sister had married James D'Arcy a century earlier. About 1718 Wyvill had purchased a prize of the wars between Austria and Turkey, the former mount of the Bashaw (commander). This horse became known as the Belgrade Turk, and his entry in the *General Stud Book* (p. 392) is copied from page lv (55) in the index to John Cheny's 1743 *Racing Calendar*. It reads in part: "Taken from the siege of Belgrade from the Bashaw of the place by General Merci who sent him to the Prince de Craon from whom he was a present to the Prince of Lorrain." This detailed information was undoubtedly supplied to Sir Marmaduke by the seller, Leopold, Duke of Lorraine (1679–1729). Cheny added that the sale was handled by the Baron de Brogue, Lorraine's diplomatic representative in London.

The Duke of Lorraine already knew Edward Coke. At age 15 in 1717, Coke had attended the academy in Luneville, capitol of the Duchy and

Left: Lexington 1850 by Edward Troye (fl. 1832–1872).
This is Boston's son, unbeaten Lexington 1850, holder of the world's record for 4 miles. He was purchased in 1850 by Robert Aitcheson Alexander of Woodburn Farm, Kentucky, where he was leading American sire for fourteen seasons, an unparalleled record.

had been entertained by the Duke. In 1727 Edward Coke inherited Longford Hall in Derbyshire, a fine estate. He had enough money to buy any horse he pleased. He founded a stud farm and began racing and breeding. Here was an opportunity for the Duke of Lorraine to sell a high-priced stallion to the racing man he had known as a 15 year old pupil at the Luneville Academy. The following account of the origin of the Godolphin is the most plausible conjecture, and the most in accord with the ample circumstantial evidence.

On July 21, 1718, Austria and Turkey had signed the Treaty of Passarowitz which restored good relations between Vienna and Constantinople. By the late 1720s, the Austrian Emperor, Charles VI, through diplomatic channels, could have easily acquired a top-class stallion from the Turkish Sultan Ahmed III. Lorraine's access to the best Turkish horses—through the Austrian Emperor—was enhanced in 1723 when his 15 year old son, Charles Steven, was placed in the court of Vienna to be groomed as a possible husband for Princess Maria Theresa, the Emperor's daughter. (They were married in 1736.) The Duke, declining in health, sent an urgent request to his kinsman, the Emperor, to acquire such a stallion.

If Coke could be induced to come to the ducal stables at Luneville, capital of the Duchy, if he could see the horse, if he could be fully informed about the fascinating background of this 5 year old stallion, he *might* be turned into a buyer. Edward Coke saw the Godolphin Arabian for the first time in France in 1729 at the Duke's stable. Lorraine's presentation of the horse and his fascinating account of its similar origins to Place's White Turk and to Coke's brother-in-law's prize of war ten years earlier was an unexpected surprise, and not to be disregarded. Edward Coke bought the Godolphin, although he already had a stallion, Whitefoot, at Longford Hall, and only five or six brood mares. The Duke died late that same year, 1729.

In 1730, "Mr. Coke is said to have imported the Arabian from France" (*General Stud Book*, Vol. I, 1808, p. 560). The Duke of Lorraine might have wished to present the 5 year old stallion as an Arabian, but Coke would not have believed a horse with such deep and sloping shoulders had pure Arabian bloodlines. Still smoldering were the political hatreds of the 1660s directed against Place's White Turk, given in 1657 to Oliver Cromwell by Mohammed IV, Sultan of Turkey. If it had been made public that Lorraine's stallion came from the same stud and stable, an outcry would have followed. Lorraine probably swallowed and told the truth, while counseling young Coke against public acknowledgment.

This was in fact an asset. The pedigree of Whitefoot 1719, the stallion Coke already owned, on the sire's side contained mostly D'Arcy of Sedbury bloodlines. His dam's pedigree traced back in tail female to a mare by Place's White Turk 1657 and also included crosses of the Helmsley Turk (flourished 1684) and of the Byerley Turk (at stud 1691–1702). This pedigree could have been approved by England's leading breeder, James D'Arcy of Sedbury (1650–1731), who crossed Place's White Turk bloodlines on Hobby and Running-Horse sprinting bloodlines. Another possible adviser was the Duke of Devonshire. In 1732, he allowed Coke to breed a mare to Flying (Devonshire) Childers 1715, a private stallion.

The Godolphin was not a chance-bred horse of mixed pedigree. The extraordinary prepotence passed on to his descendants could only have been based on a strain nurtured for centuries by breeders whose consistent selections of breeding stock were based on a fixed standard of conformation and performance. The only 'superior speed' strain which could fulfill these specifications was the Turcoman-Arabian diplomatic gift strain of the Grand Signors of the Ottoman Empire.

Who First Called the Godolphin an Arabian?

Neither the description of the Godolphin Arabian by Osmer (1756) nor the portraits by Morier (1753), copied by Stubbs and Sartorius, bear any resemblance to purebred Arabians, such as the Darley Arabian 1704 or the Oxford Dun Arabian 1715. Yet the only way Coke could charge a profitable stud fee was to call the stallion an Arabian. The truth that he was a Turcoman horse might have attracted some curiosity, but it would not have been good for business.

For example, a newspaper advertisement in *The Mercury* of May 19, 1746, published in Stamford, Leicestershire reads:

> Now in the hands of Mr. Bakewell of Dishley, a fine, brown-bay stallion of the Turcoman breed, fifteen hands and one inch high, very strong, bought out of the Grand Signor's stud [Sultan of Turkey]. One Guinea the mare and two shillings and sixpence the servant. (Thomas Ryder, former editor of the *U. S. Carriage Journal* in a letter dated September 24, 1994.)

Robert Bakewell (1725–1795) was world-famous for his Leicester rams, and as a pioneer in improved livestock breeding. Yet in spite of the stallion's royal origin and the well-known farm where he stood, his identity as a Turcoman horse reduced the stud fee to "One Guinea each mare and two shillings and sixpence the servant."

There is another plausible answer to the question of how the Godolphin came to be called Arabian. Gifts of horses by sovereigns to other heads of state were an important part of mid-Eastern diplomacy. Because of special requirements, diplomatic gift horses were a separate strain. They *had* to be beautiful. In her weighty volume *Thoroughbred Racing Stock* (1938), Lady Wentworth cites sixteenth and seventeenth century texts that the Emperors of Morocco used part-Arabian and part-Egyptian horses. Because purebred Arabians were considered national treasures and their export strictly forbidden, the Sultans of Turkey could not present them to other heads of state as diplomatic gifts. However, in their stables, available for cross-breeding, they had both Turcoman and Arabian horses, the most beautiful in the world.

There is graphic evidence that the pedigree of the Turcoman stallion known as the Godolphin Arabian contains one or more crosses of Arabian blood in Shaw's portrait of King's Son of Blank 1764. The portrait of this horse, whose sire and dam were both by the Godolphin, shows the beautiful head and high-set tail of the purebred Arabian, and the deep sloping shoulders of the Turcoman.

At the beginning of Part Four of the *General Stud Book*, headed "Arabians, Barbs and Turks," it is explained (p.388) that "some were not imported at all—it being common [practice] to call horses Barbs, Turks, etc. [Arabian] that were only of that breed." In other words, that "had an imported [Barb, Turk, etc.] ancestor." Quite fairly, Edward Coke, his importer, called him an Arabian. Furthermore, since this Turcoman stallion had an Arabian ancestor, the Godolphin Arabian was fully qualified for registration in the 1791 *General Stud Book*.

What's in a Name?

It was extraordinarily fortunate for the breed that Coke called his stallion an Arabian. If the inventory of Coke's estate had listed the newly-imported stallion as a Turcoman horse, subsequent Thoroughbred history might have been very different. Godolphin would not have been interested in acquiring, either by purchase or by gift, a plain strange-looking horse of an unpopular breed, with no pedigree, no racing record and no winning offspring. Williams would then have sold him at a low price to cover farmers' mares. Marske (sire of Eclipse) had this experience. Had Coke not called his horse an Arabian, and had Lord Godolphin not given the young stallion his opportunity as a sire, he probably would not have influenced the Thoroughbred breed as he did. Nor would he have been included in the *General Stud Book*.

CHAPTER XI

The Duke of Cumberland and Eclipse

H.R.H. the Duke of Cumberland (1721–1765)

During the middle of the eighteenth century, the dominant figure in British racing, and by far the most successful breeder of Thoroughbreds, was H.R.H. the Duke of Cumberland, second surviving son of George II. Born April 15, 1721, he was elevated to the Dukedom in July 1726 when he was 5 years old.

Military Career

Cumberland's youthful ambition was to be a famous general in military history. In April 1740, at age 19, he was appointed Colonel of the Coldstream Guards and 2 years later, Major General. Since the Kings of England were also rulers of Hannover in Germany, British troops were sent to fight in the War of the Spanish Succession. In the Battle of Dettingen (June 27, 1745) where he fought "with conspicuous bravery," Cumberland received a leg wound which never healed properly, and which kept him from riding, hastening his death at the age of 44 in 1765.

On April 16, 1746, defeating the Scottish invaders at Culloden, Cumberland became a national hero. In 1747, however, returning to the continent in the Seven Years War, his fortunes changed. On September 8, opposed by the superior forces of Marshall Saxe, and trapped between the North Sea and the River Elbe, Cumberland was forced to sign the Covenant of Kloster Seven, later repudiated by the British Parliament.

After Culloden, the King called Cumberland "the best son that ever lived," but a year later, after Kloster Seven, he said "my son has ruined me and disgraced himself."

Cumberland Lodge

Fortunately for the future of the Thoroughbred, the Duke of Cumberland was so hurt and humiliated by these events that he gave up his military career and returned to his other love—racing and breeding horses. On July 12, 1745, he had been appointed Ranger of the Windsor Great Park. In the autumn of 1747, when he was 26 years old, Cumberland returned from the scene of his disgrace and moved to the Windsor Park House, ever since known as Cumberland Lodge. His additions to the lodge are said to have included a picture gallery in which portraits of famous racehorses were hung. His brood mares, stallions, racehorses and young stock were kept at the adjacent Cranbourne Lodge.

Leader of British Racing (c. 1752–1765)

The disappointments which ended Cumberland's military career had little effect on his eminence among members of the "sporting aggregate." He was one of the group whose meetings in 1752 led to the formation of the Jockey Club, of which he was a founding member. Cumberland became in effect the leader of English racing.

Arabian Bloodlines Experiments

Cumberland helped to direct the experiments leading to the conclusion that line breeding to Arabian strains was producing slower horses racing over 4-mile courses. These conclusions were described in the *Sporting Dictionary* (Vol. II, p. 208) by William Taplin, who added, "This discovery having been made ... at the very moment of the Great Duke of Cumberland having brought the sport of racing to nearly its zenith of attraction and celebrity."

Cumberland believed, correctly, the Godolphin Arabian was the most influential sire of the eighteenth century. He also knew that the horse was not an Arabian.

Racing and Breeding

From 1748 until his death in 1765, the Duke of Cumberland was a constant and familiar figure at British race meetings. He maintained a very active stable of racehorses, most of them home-breds sired by sons and grandsons of the Godolphin Arabian. Many of these were winners,

The Duke of Cumberland and Eclipse

notably King Herod 1758, who got Highflyer 1774, winner of the most purse money in the eighteenth century. The Duke of Cumberland's exalted reputation in the history of the Thoroughbred, however, is due to his remarkable success as a breeder.

The Great Progenitors

During the eighteenth century the attention of racing men was focused primarily on male bloodlines, specifically on the three sires known as "The Great Progenitors." These were the stallions popularly believed to be responsible for continuing the three Thoroughbred tail male lines. The line of the Byerley Turk (went to stud in 1690 or 1691) was carried on by King Herod 1758. The line of the Darley Arabian (imported 1704) was carried on by Eclipse (foaled in 1764). The line of the Godolphin Arabian (imported in 1730) was carried on by his grandson Matchem (foaled in 1748).

King Herod and Eclipse were both bred by the Duke of Cumberland. He was aware that during the latter part of the seventeenth century, the 4-mile heat King's Plate courses, founded by Charles II in 1665, had precipitated the decline of sprinting Running-Horse and Hobby bloodlines. He knew that only a few long-established breeders continued to preserve the female lines essential to maintaining these short speed strains.

Profound Knowledge of Bloodlines

The Duke of Cumberland was not a long-established breeder. Nevertheless, 40 years before the publication of the *General Stud Book*, he had acquired an encyclopedic knowledge of pedigrees and the various strains which they included. Furthermore, he had a clear understanding of both the strengths and weaknesses of the individual strains. This was illustrated by his contributing to the 1753 and c. 1759 inscriptions accompanying the Morier portrait of the Godolphin Arabian (see Chapter X). Cumberland's contributions are the first and the most perceptive explanations of the Godolphin's preeminence as a sire.

Optimum Combination of Speed Strains

For Cumberland, one question remained. In order to produce superior race horses, which strains should be used and in what combinations? He was an experimental breeder. His disappointment with crossing Arabian strains has been mentioned. Cumberland believed additional experiments in cross-breeding could provide answers to his question.

He utilized the following strains in his experiments: The first strain was the short speed Hobby strain bred by the Earls of Rutland during the sixteenth and seventeenth centuries at the Helmsley stud. The second strain was the short speed Running-Horse strain bred at Tutbury by Charles I (died 1649) and the "Royal Mares" acquired by the D'Arcys of Sedbury (c. 1650–1731). Third was the middle distance speed strain carried by the descendants of the Godolphin Arabian and Place's White Turk, the Turcoman-Arabian sires imported in 1730 and 1657.

Experiment # 1: The Prepotency of the Godolphin Arabian

In 1756, the Duke of Cumberland purchased the bay unraced 6 year old mare Cypron (Blaze ex Selima by Bethell's Arabian) from Sir William St. Quinton of Scampston, North Yorkshire. Cypron's pedigree, in tail female, traced back to an obscure "daughter of Old Merlin" (Bruce Lowe #26). Seven-eighths of the pedigree is undistinguished. Her foals were replicas of their sires. Cypron thus provided the right kind of foundation for experiments based on breeding her to a selected sire with a bloodline carrying one of the three leading speed strains.

Testing the prepotency of the Godolphin Arabian had already begun. In the inscription proposed for the unpublished 1759 copy of the 1753 Godolphin Arabian print, Cumberland praised the stallion for "hitting with most other pedigrees," good, bad and indifferent. Cypron's pedigree was indifferent. She was in foal to Regulus when he bought her. Cumberland also purchased her weanling, Dumplin, and her yearling, Dapper, both by Cade, who, like Regulus, were sons of the Godolphin Arabian. Here were three experiments, already in progress.

The Godolphin lived up to Cumberland's assessment. Dumplin was a good racehorse, and Dapper 1755 was probably the best 4-year-old of his year. (The birthday of racehorses was then May 1st, not January 1st.) Dapper, for the next 2 years, "was used as a trial horse," in private, to gallop against other horses in order to test their condition and speed (Pick, *Turf Register*, 1803, pp. 251, 231).

Throughout his life the Duke of Cumberland believed, correctly, that the Godolphin Arabian was England's leading sire. This belief was confirmed by the excellent race record as a 4-year-old compiled in 1759 by the Godolphin Arabian's grandson, Dapper, out of the undistinguished mare Cypron.

Experiment #2: King Herod

The Duke of Cumberland was apparently convinced that Place's White Turk, a Turcoman-Arabian stallion, was the tail male ancestor of the Byerley Turk. A registered tail male descendant of the latter was Tartar, foaled in 1743, bred by Robert Leedes of North Milford, near Tadcaster, a stud established circa 1668 by his father Edward with two foundation mares, granddaughters of Helmsley's Hobby strain Old Bald Peg.

Herod 1758, bred by the Duke of Cumberland, was by Tartar 1743, a tail male descendant of the Byerley Turk. The Duke was convinced that the Byerley Turk was a tail male descendant of Place's White Turk (G.S.B., p.388), the seventeenth century source of Thoroughbred middle distance (4-mile course) speed. Herod was an excellent 4-mile King's Plate racehorse. The winnings of his get exceeded those of any other eighteenth century stallion. Although it was more than 100 years since Place's White Turk landed, his blood still produced middle distance speed in Herod and in Herod's unbeaten son, Highflyer 1774.

Speed and the Thoroughbred

Highflyer by John Boultbee.
Highflyer Hall is in the background. Richard Tattersall named the Hall in honor of his horse.

Place's White Turk 1657

Cumberland wished to test the prepotency of Place's White Turk, although the line was by then 100 years old. He bred Cypron to Tartar, a mating with no Godolphin ancestry. William Pick in his 1803 *Turf Register* (p. 99) noted that Tartar "was nigh [just under] 15 hands, a horse of great strength and power." These words also describe the conformation of a seventeenth century sprinter.

Cypron's 1758 foal by Tartar was the bay horse King Herod. By the spring of 1763 Cumberland was ready to assess the result of the experiment. The Godolphin's grandson, Dapper 1755, the best four-year-old of

The Duke of Cumberland and Eclipse

Richard Tattersall by Thomas Beach.
Racehorse auctioneer Richard Tattersall owned Highflyer, whose portrait is on the previous page. Note the painting of Highflyer in the Tattersall portrait.

his year, had become the stable's trial horse. Probably to Cumberland's surprise, in works against Dapper over the home gallops, King Herod proved to be the better racehorse.

In his 1803 *Turf Register* speaking of King Herod, Pick said he was "allowed to be one of the best bred horses produced in this kingdom." He was a top-class racehorse. At stud the total racecourse winnings of his offspring were larger than those of any other eighteenth century sire, including Eclipse. A major contributor to this total was his son, the unbeaten Highflyer 1774, whose dam, Rachel, had two crosses of the Godolphin in her pedigree. Highflyer's stud fees amassed "a noble for-

Speed and the Thoroughbred

tune" for the racehorse auctioneer Richard Tattersall, and allowed him to transform Highflyer Hall into a magnificent mansion.

This tail male line is the second longest in the Stud Book, of roughly 250 years. It extended from imported Place's White Turk through the Byerley Turk, Tartar, King Herod, Highflyer, to The Tetrarch 1911, a top-class 2-year-old and sire of sprinters. He, in turn, was the sire of Mumtaz Mahal 1921, granddam of Nasrullah 1940, the sire of Bold Ruler 1954 and grandsire of Secretariat 1970.

Experiment #3: Eclipse 1764

The year before his death in the autumn of 1765 the Duke of Cumberland carried out a third experiment. His primary interests were in pedigrees, in bloodlines, in how to produce maximum speed by crossbreeding the Thoroughbred's three source strains: the sprinting pre-Christian Irish Hobby, the sixteenth century English Running-Horse, and the imported middle distance Turcoman-Arabian strain.

Cumberland believed that British breeders were paying too much attention to the racing records of sire and dam, and not enough attention to the record at stud and on the racecourse of earlier generations. If such an unorthodox mating should produce an outstanding racehorse, Cumberland would prove his point.

Fortunately, the Duke of Cumberland was more interested in the best combination of speed strains than in protecting the reputation of the Godolphin Arabian. A further experiment was needed, the sooner the better. It began in May of 1763. Cumberland went no farther than his own bloodstock in the paddocks of adjacent Cranbourne Lodge.

He had a 13 year old stallion, Marske, whose sire had narrowly escaped being fed to the hounds. Cumberland had acquired this horse as a weanling from John Hutton III by trading an Arabian stallion, because the third, fourth and earlier generations of his pedigree included a bountiful heritage of seventeenth century sprinting speed strains. Because of his modest racing record, Marske had been bred only to farmers' mares at a guinea apiece.

Cumberland also had an unraced 14 year old mare, Spiletta, who previously had produced only one foal, a 4 year old filly, also unraced. The earlier generations of her pedigree contained many seventeenth century sprinters. Most importantly, her sire was the Godolphin Arabian. The Marske-Spiletta mating produced a colt foal in the spring of 1764, according to the *General Stud Book* (p. 197).

The credentials of sire and dam did not impress racing men in the eighteenth century any more than they would today. At the dispersal sale held a year after the Duke's death in 1675, organized by Richard Tattersall, the colt (then a yearling) was sold for 75 guineas, a low price based solely on his good looks. Since he was born on April 1, 1764, a day when the sun was almost totally darkened by the moon, his new owner, William Wildman, a grazier and Smithfield meat merchant, called the colt Eclipse.

Eclipse grew up to become an unbeaten race horse and leading sire of racehorses in the eighteenth century. The Duke of Cumberland died on October 31, 1765, when Eclipse was a year and seven months old. Cumberland did not live to see his unbeaten colt win one 4-mile heat race after another in 1769 and 1770. Had he lived longer he would not only have enjoyed the vindication of his theories as a breeder, he would also have received the congratulations of the entire world of horse racing.

In April 1764, the Duke of Cumberland probably never dreamed that the bloodlines of his little chestnut foal would fortify and establish the tail male line of its great-great-grandfather, the Darley Arabian, the male line of approximately 90% of all the world's Thoroughbreds today, the only male racing line which survives unbroken since the eighteenth century.

In his entertaining and beautifully illustrated book, *Eclipse and O'Kelly* (1907, p. 44), Sir Theodore Cook praised the Duke of Cumberland as follows:

> [T]o him belonged *Herod* and *Eclipse,* two out of those three great sires from whom all our racehorses have since descended, and that he owned *Eclipse's* sire and dam, and the dam of *Herod* as well. No man can fairly be said to have done more for English Racing.

"The Rest Nowhere"

Eclipse made his first start in the *Noblemen's and Gentlemen's Plate* at Epsom on May 3, 1769. Cook described the circumstances as follows:

> Word got out that Mr. Wildman had a promising five year old entered on May 3 (1769) in a 250 guinea plate at Epsom, for horses that had never won £30. It was also rumoured that he would be tried against a good horse on the downs the morning before race day. Betting touts arrived too late for the trial, but an old woman said 'she had just seen a horse with a white leg running away at a monstrous rate and another horse a long way behind trying to race after him, but she was sure he would never catch the white-legged one if he ran to the world's end.'

Speed and the Thoroughbred

When race-goers gathered at Medley's Coffee House that afternoon, Dennis O'Kelly, a shrewd Irish professional gambler, and his friends got their bets down.

On race day, ridden by John Oakley, Eclipse easily beat the other four starters in the first 4-mile heat. Cook wrote:

> Desirous of adding to his gains, and after the first heat, being perfectly confident that this great horse could race as well as he could gallop, O'Kelly made a heavy wager (which was taken up with considerable eagerness), that he would place the five horses in the second heat. When asked to name their order he pronounced his famous sentence 'Eclipse first, the rest nowhere,' as he was sure that all the others would be 'distanced' (to be distanced was to be beaten by over 200 yards) and therefore would not be placed by the judge. John Oakley, his jockey, only had to sit quite still and though all the horses were bunched close together at the 3-mile post, *Eclipse* sailed away so easily from there that he beat the rest, hard held, by more than the required distance. His jockey could not have stopped him if he had tried.

O'Kelly's fortune was made.

Eclipse and His Owners

Eclipse was the most famous racehorse of the latter half of eighteenth century and its most influential sire. His pedigree descended from the four eminent studs—Helmsley, Tutbury, Wallington and Welbeck—the latter three of which were casualties of the English civil wars (1642–1660). Helmsley stud was undamaged, presented to Lord Fairfax by the new Parliament, and later inherited by his daughter who married the Second Duke of Buckingham. James D'Arcy of Sedbury stud had reassembled the brood mares of the other three studs. Miraculously, the breed survived.

The first owner of Eclipse was royalty, the Duke of Cumberland, second surviving son of King George II. Cumberland's choice of brood stock—Marske, servicing farmers' mares at a guinea apiece and unraced Spiletta with one unraced filly—was virtually a slap in the face, however, to the aristocratic breeders who owned the most popular stallions. At the Duke's dispersal sale following his death, the handsome yearling was purchased for only 75 guineas by a merchant, William Wildman. The next owner was even more embarrassingly distant from the Royal throne, in the person of Dennis O'Kelly, the Irish professional gambler! That this

The Duke of Cumberland and Eclipse

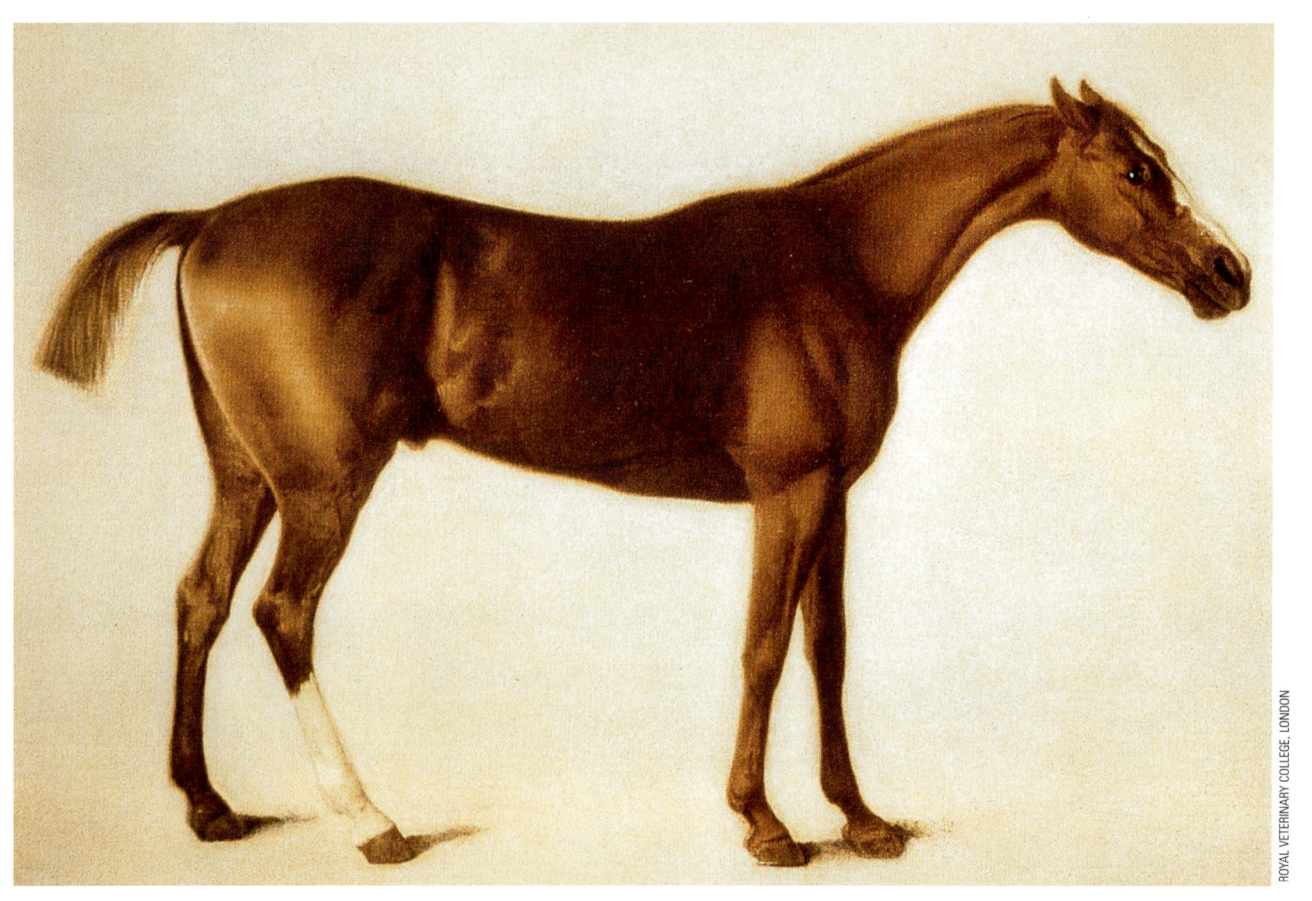

Eclipse by George Stubbs (1724–1806), sketch in oils.
Before beginning his more formal portraits of Eclipse, George Stubbs made a sketch of the horse from life. It shows conformation similar to that of many modern Thoroughbreds.

descent into unfashionable ownership does not seem to have adversely affected his career at stud is a striking tribute to Eclipse's greatness as a sire.

The conformation of Eclipse became the classic model for Thoroughbred conformation, not only for those who saw the horse in life, but also for those who later were able to look at the several portraits painted from life by George Stubbs. Notable is the sketch in color of Eclipse in racing condition, presumably painted in the autumn of 1770, before he had added the extra weight of a stallion standing at stud.

CHAPTER XII

Surviving Speed Lines into the Twentieth Century

Sources of Thoroughbred Speed

The three racing speed sources of the sixteenth, seventeenth and eighteenth centuries have been preserved and passed on through tail female bloodlines, not through tail male bloodlines.

For many past centuries the sheiks of the Syrian desert have recited and thus preserved the extended pedigrees of their Arabian horses, the most beautiful of all breeds. These horses are also the fastest over long distance—some strains having been developed over centuries of desert warfare, other strains having been bred for racing for 2,000 years and more. These pedigrees contain only tail female lines. Stallions are considered unimportant. This had been true in all horse-racing oriented cultures since antiquity. The quarter-mile Hobby and Running-Horse mares up to the sixteenth century were cherished as the repositories of sprinting speed. Since then, however, the breeders of Thoroughbred racehorses placed primary emphasis on male lines, on stallions with tail female sprinting Hobby and Running-Horse strains—lines which have lasted a generation or so and then faded.

King's Plates 1665

King Charles II abruptly changed the principal pattern of English horse racing in 1665. He founded at Newmarket the first of a series of King's Plates awards, gradually increased to include more than twenty of the principal racing centers in England and Scotland. The King's *"Articles of Racing"* specified two to four heats over 4-mile round courses. This required middle distance speed, for multiple heats, a quality that had not, until then, been required to breed winning racehorses in England.

Two imported stallions, from the strain developed by the sultans of Turkey as diplomatic gifts, supplied the required middle distance speed. This strain was based on Turcoman stallions and mares, their rather plain looks enhanced by crosses of the prized and protected pure-bred Arabian bloodlines. The Turcoman-Arabian sires who completely altered the racehorse breed were Place's White Turk, imported in 1657 as a gift from Sultan Mohammed IV to Oliver Cromwell and the Godolphin Arabian, imported by Edward Coke in 1730, 73 years later. These two stallions were exceptional in their influence. When crossed with the sprinting speed strains, the result was a horse who could run the middle distances with speed.

Place's White Turk, the first supplier of the desperately needed middle distance speed, was already in England in 1665. He had survived the upheaval of Cromwell's death and the Restoration due to the devotion of his handler, who, without permission, had risked moving the stallion to the safety of his family's seat Dinsdale, close to the top sprinting strain mares of North Yorkshire. In almost fairy-tale fashion, Place's White Turk was a stallion in his prime at the very moment that the King created a need for exactly what this white stallion had to offer, the third and final source of speed. As the century closed, his bloodlines had saturated the breed.

Three decades later, in 1730, it was the Godolphin Arabian, the second Turcoman-Arabian import, who refreshed the middle distance speed strain, and changed the breed into the horse we know today. The Godolphin gave the breed its sloping shoulders, deep chest and high hindquarters.

Place's White Turk and the Godolphin Arabian founded the tail male bloodlines of the Thoroughbred breed during the seventeenth, eighteenth and nineteenth centuries. Their early careers at stud were highly successful, but they did not endure.

The tail male line of Place's White Turk was carried only by the Byerley Turk (at stud c. 1690), King Herod 1758 and Highflyer 1774, who was the largest purse winner of the eighteenth century. After modest careers in England and France during the nineteenth century, the White Turk's tail male line ended in glory with the champion 2 year old racehorse and sire The Tetrarch 1911. His spectacular speed has been carried on through his daughter, the famous brood mare Mumtaz Mahal 1921.

Perhaps because of an excessive concentration of middle distance speed bloodlines, the length of the tail male line of the Godolphin Arabian 1730 was even shorter. The last top-class sire in this line was West Australian 1850, winner of the Epsom Derby and the Doncaster St. Leger and the sire of many good racehorses. However, none of West Australian's sons were able to reproduce their sire's record at stud.

Finally, there is the tail male line of the Darley Arabian 1704, the only such line of racing Thoroughbreds which survives unbroken since the eighteenth century. Approximately 90% of all present day Thoroughbreds descend tail male from the Darley. The problem is, the Darley was an authentic Arabian with no middle distance speed. The mares to which the sires of this line were bred had pedigrees which were plentifully endowed with both middle distance speed *and* the dazzling sprinting speed of the Hobby and Running-Horse mares.

Periodic infusions of sprinting bloodlines are needed to refresh middle distance speeds. Notable among the mares that furnished these infusions are Spiletta by the Godolphin Arabian, dam of Eclipse 1764, and Anthony Childers' mare Betty Leedes, bred to the Darley Arabian. She produced Flying Childers 1715, called Devonshire Childers in the *General Stud Book*, and his full brother Bartlett's (Bleeding) Childers 1716, tail male great-grandsire of Eclipse. The pedigree of Betty Leedes contains three crosses of the famous sprinting speed strain mare Old Bald Peg c. 1635. Although the Darley Arabian did not contribute speed to the Thoroughbred, he did contribute his spirited temperament and his beauty. This contribution was made through these two sons who were enormously popular sires. The speed in this so-called tail male line is feminine, passed on through their dams, not their sires.

In the twentieth century, tail male lines of more than two or three generations have become relics of the past. It is rare that a top-class sire has a son who is an equally top-class sire. A top-class grandson is very rare indeed.

Nineteenth and Twentieth Century Thoroughbred Speed Sires

The Duke of Cumberland, and other mid-eighteenth century commentators, noted the Godolphin Arabian (1724–1753) had "refreshed the English blood more than any other foreign horse, before or since." To the sprinting speeds of the ancient Irish Hobby and the sixteenth century

English Running-Horse, Place's White Turk 1657 and the Godolphin added middle distance. This middle distance speed is considered "speed over a distance of ground," which in the seventeenth and eighteenth centuries was over 4-mile courses, and into the nineteenth and twentieth centuries was over mile-and-a-half courses.

Middle distance speed is not self-perpetuating. By 1880 the pattern—breed an Oaks winning mare to a Derby winning sire—was wearing thin. The Thoroughbred breed, to be refreshed, needed stallions dominant for sprinting speed. Sedbury, the last great stud to breed sprinters, had been dispersed after the death, in 1731, of its owner, Lord D'Arcy of Navan. The last preeminent eighteenth century sire dominant for sprinting speed only was (Little) Janus (1746–1780), imported into Virginia in 1756. It would be hard to write a better middle distance pedigree than that of Janus (the top three quarters has survived) which closely parallels the pedigree of Eclipse. Despite his pedigree, although Janus regularly served over 300 mares a (long!) season, none of his foals could carry their blazing speed for more than 500 yards!

Importance of Sprinting Speed

One of the earliest nineteenth century breeders to realize the importance of sprinting speed was Viscount Falmouth of the Mereworth Castle stud (A. S. Hewitt, *The Great Breeders and Their Methods*, 1982). In 22 years, Falmouth bred nineteen winners of the five English classic races for 3-year-olds (Derby, Oaks, etc.). Of the twenty-four brood mares at Mereworth in 1880, twenty-two had won races as 2-year-olds, most of these at sprinting distances.

Six Surprise Sires 1881–1915 Dominant for Sprinting Speed

During the late nineteenth and early twentieth centuries, Thoroughbred speed bloodlines were refreshed primarily by six sprinting sires who emerged apparently by chance. It has long been recognized that livestock breeding is not a science but an art. The art of the Thoroughbred breeder, from Lexington 1850 to Secretariat 1970, has been set forth brilliantly in detail by Abram S. Hewitt in *The Great Breeders and Their Methods* (1982). The pedigrees and racing records of these six stallions had little in common except sprinting speed. Their places of birth, breeders, owners, and trainers were far apart. Their training methods differed widely. These stallions mostly sired very fast horses over short courses. Strangely, in one way or another, they were all surprises.

St. Simon 1881, bred by the Duke of Portland at Welbeck Abbey, compiled a magnificent record both as a racehorse and as a sire.

St. Simon 1881
The first (and greatest) was St. Simon 1881, bred by the Duke of Portland at Welbeck Abbey. Here, in the early decades of the seventeenth century, the Duke's ancestor, the First Duke of Newcastle, bred Running-Horse sprinters of which he wrote, "I have ridden many hundreds of matches" (1667). St. Simon was so little regarded that he was not entered in the classic races for 3-year-olds. When he was unbeaten in two seasons of racing, His Grace was surprised. When he compiled a record as a sire (which Hewitt considers the greatest in Thoroughbred history), doubtless His Grace was again surprised—and certainly gratified! St. Simon was dominant for sprinting speed and for all other speeds.

Domino 1891

Before the Barak Thomas dispersal sale to which Domino was consigned as a yearling in 1892, James R. Keene of Castleton Stud, Lexington, Kentucky, had decided not to bid. He was surprised to learn, after the sale, that the colt had been bought for his account by his son, Foxhall Keene. Domino and his son, Commando, both short-lived, were major contributors to the sprinting speed of the American Thoroughbred.

Barak Thomas, breeder of Domino, bought Keene Richards' collection of the superb portraits of leading nineteenth century Thoroughbreds painted by Edward Troye (fl. 1832–1872) from Richards' widow. These portraits now adorn the Boardroom of the Jockey Club (American).

Americus 1892

It was a surprise that Americus, bred in California by "Lucky" Baldwin in 1892, was acquired by Richard Croker, "Boss" of Tammany Hall. It was more surprising that about 1901, after New York politics got too hot, Americus was moved to Ireland. Had his daughter of 1905, Americus Girl, been foaled 8 years later (1913), the provisions of the infamous "Jersey Act" would have branded her a half-bred, ineligible for the *General Stud Book*. That would have eliminated her descendants, Mumtaz Mahal 1921, *Nasrullah 1940, Bold Ruler 1954, Secretariat 1970 and many other famous horses.

The Tetrarch 1911

In order to strengthen the tail male line of King Herod 1758 in Ireland, Edward Kennedy imported Roi Herode from France. His surprising son, The Tetrarch 1911, unbeaten at two, was the father of the good racehorse and sire Tetratema and of the very fast filly Mumtaz Mahal 1921, granddaughter of Americus Girl 1905.

Phalaris 1913

Phalaris was the fastest horse of his day, a sprinter, whose distance capacity was less than a mile. During World War I, Lord Derby tried to sell him for £5000. Fortunately there were no takers. Phalaris supplied the Derby stud with the sprinting speed in which it was previously deficient. His descendants in tail mail were Pharos, *Nearco, *Nasrullah, Bold Ruler and Secretariat. This sequence, presumably, was a surprise to Lord Derby.

Surviving Speed Lines into the Twentieth Century

The Tetrarch in 1914 by A.G. Haigh

Havresac II 1915
Esteemed as the most brilliant breeder of this century, Federico Tesio maintained at his Dormello stud, near the Italian Alps, several brood mares with modest credentials, the best he could afford. His mares had access to only a few top-class proven sires in England. Among local sires he tried Havresac II 1915, leading sire in Italy, where the total annual crop of Thoroughbred foals was about 300.

At his Dormello stud, to the cover of this previously obscure sire, Tesio bred father and son "Horses of the Century," Nearco 1935 and Ribot 1952. Tesio was undoubtedly surprised—pleasantly. He first bred to Havresac II in 1925. Hewitt (p. 384) wrote: "Tesio could hardly have known at that time that Havresac II was dominant for [sprinting] speed."

The high hindquarters, which the Godolphin Arabian passed on to his late nineteenth and twentieth century descendants, seem to be associated with horses dominant for speed. This is confirmed by photographs of St. Simon, The Tetrarch, Phalaris, and High Time.

The concentration of sprinting bloodlines in the extended pedigrees of our most eminent sires and brood mares is remarkable. The six-generation pedigree of Bold Ruler 1954 (sire of Secretariat 1970) contains all six of these sires dominant for sprinting speed.

Fifteen Late Seventeenth Century Sprinting Speed Tail Female Families

Most early British breeders paid great attention to tail male bloodlines. Only a few gave weight to female lines. It was not until 1895, when C. Bruce Lowe identified and ranked 43 female lines based on performance, that a definitive work on female bloodlines appeared in print. Lowe's text identifies the taproot mares who are the most important late seventeenth century sprinting speed strain mares. Of the first fifteen families, two are from the Rutland's Helmsley stud, and at least nine are mares salvaged by the D'Arcys after England's bitter civil wars (1642–1660). Four are registered as Royal Mares, from the Tutbury Royal stud of Charles I (r. 1625–1649).

The Thoroughbred breed has no surviving middle distance speed tail male bloodlines. It does have fifteen late seventeenth century sprinting speed tail female lines still to be found in the pedigrees of our most eminent twentieth century racehorses and sires. As an example, consider the top-class stallion Bold Ruler 1954, sire of Secretariat. *Of the 64 ancestors in the sixth generation of Bold Ruler's pedigree, it is remarkable that 63 trace back in tail female to one or another of the late seventeenth century taproot mares ranked 1–15 by C. Bruce Lowe,* Breeding Racehorses by the Figure System. (Illustration of the pedigree is courtesy of the Keeneland Library, Lexington, Kentucky.)

Bold Ruler 1954 #8	Nasrullah 1940 #9	Nearco 1935 #4	Pharos 1920 #13	Phalaris 1913	#1	Polymelus 1902	#3
						Bromus 1905	#1
				Scapa Flow 1914	#13	Chaucer 1900	#1
						Anchora 1905	#13
			Nogara 1928 #4	Havresac 1915	#8	Rabelais 1900	#14
						Hors Concours 1906	#8
				Catnip 1910	#4	Spearmint 1903	#1
						Sibola 1896	#4
		Mumtaz Begum 1932 #9	Blenheim II 1927 #1	Blandford 1919	#3	Swynford 1907	#1
						Blanche 1912	#3
				Malva 1919	#1	Charles O'Malley 1907	#5
						Wild Arum 1911	#1
			Mumtaz Mahal 1921 #9	The Tetrarch 1911	#2	Roi Herode 1904	#1
						Vahren 1897	#2
				Lady Josephine 1912	#9	Sundridge 1898	#2
						Americus Girl 1905	#9
	Miss Disco 1944 #8	Discovery 1931 #6 (#23)	Display 1923 #2	Fair Play 1905	#9	Hastings 1893	#21
						Fairy Gold 1896	#9
				Cicuta 1919	#2	Nassovian 1913	#14
						Hemlock 1913	#2
			Adriane 1926 #6 #23	Light Brigade 1910	#8	Picton 1903	#7
						Bridge of Sighs 1905	#8
				Adrienne 1919	#6 (#23)	His Majesty 1910	#6 (#23)
						Adriana 1905	#6 (#23)
		Outdone 1936 #8	Pompey 1923 #3	Sun Briar 1915	#8	Sundridge 1898	#2
						Sweet Briar II 1908	#8
				Cleopatra 1917	#3	Corcyra 1911	#6
						Gallice 1910	#3
			Sweep Out 1926 #8	Sweep On 1916	#3	Sweep 1907	#8
						Yodler 1899	#3
				Dugout 1922	#8	Under Fire 1916	#9
						Cloak 1909	#8

The 4-digit numbers are foaling dates. Of the 64 horses in the pedigree's 6th generation (not shown), 63 trace back in tale female to Bruce Lowe's family 17th century foundation mares numbered 1–14.

Speed and the Thoroughbred

The world's racing countries have enjoyed racing in the English manner for several centuries. Today we have the benefit of an accumulated wealth of knowledge of pedigrees and results of well-chronicled information, time-keepers, daily newspapers, instant communication and computers. But we still have a lot to learn about the major importance of the tail female lines.

Summary

Speed and the Thoroughbred has chronicled the evolution of the Thoroughbred, exploding some myths, uncovering the Turcoman influence, and tracing the tail female speed strains which preserved the breed's sprinting speed, passing it on from generation to generation. Over time, the Thoroughbred breed suffered from the ravages of war and from political and social intrigues, but flourished under the care and attention of nobility and wealthy landowners. Thanks to a few visionary breeders, the Thoroughbred ultimately evolved to become the fastest horse in the world.

This evolution is marked in equal measure by careful planning and chance events. Of the three major sources of speed, the first was the speedy Irish Hobby, introduced into the studs of major racehorse breeders in England in the early 1600s, providing sprinting speed strain mares who were the foundation of the English racehorse known as the Running-Horse. These Running-Horse mares were the second source of speed. The Hobby and Running-Horse speed strain mares provided the source of speed that flourished under James I and Charles I, who patronized sprint racing and breeders, and maintained their own studs, until England's civil wars.

The civil wars might have caused the breed to disappear, had it not been for several fortuitous events. The first saviors of the breed were the D'Arcys, who quietly reassembled, before the Restoration of Charles II, the best sprinting mares in England after they had been scattered during and after the war. Ironically, when the civil wars ended, it was Fairfax, recipient of the Helmsley stud as a spoil of that war, who made another important contribution to the breed's recovery. He bred the Old Morocco Mare 1654, a daughter of the Old Morocco Barb, probably a 1637 gift by the Emperor of Morocco to Charles I. She was out of the foundation Hobby strain mare Old Bald Peg. The two major sources of speed had been recovered and were in the hands of able men.

With the introduction of King's Plates in 1665, decades of sprinting speed strains were faced with the necessity of adapting to racing over multiple 4-mile heats. Place's White Turk, the Turcoman-Arabian sire taken to safety by Rowland Place after the collapse of Oliver Cromwell's Parliament, was well-located to introduce the necessary middle distance speed to the breed. He provided the third and final source of speed.

Place's White Turk was undeniably the most influential sire prior to the Godolphin (Turcoman-)Arabian, who was imported 73 years later. Both Turcoman-Arabian stallions brought middle distance speed, but it was the Godolphin's prepotence that transformed the conformation of the breed into the horse we recognize as the Thoroughbred today.

Despite the different influences of many important sires, the tail female lines continue to supply the breed with the speed strains making the Thoroughbred the world's fastest horse. We can consider the female lines as the repository and preservers of these speed strains. Of all the tail male lines, only one racing line survives unbroken, that of the Darley Arabian. Yet, remarkably, fifteen sprinting speed tail female lines from over three centuries ago are found today in the pedigrees of our most eminent racehorses and sires.

BIBLIOGRAPHY

Book of Leinster, Fulartach, c. 1000–1047

Calendar of State Papers, Milan, 1385–1618

Harleian Manuscript, 1388

The History of Greater Britain, John Major, translated from the 1522 Latin editions by A. Constable, Edinburgh, 1882

Historica Descriptio Hiberniae, Paulus Jovius, 1548

Gli Ordini di Cavalcare, Federico Grisone, 1550, translated by Thomas Blundeville, 1560

The foure chiefest Offices belonging to Horsemanship, Thomas Blundeville, London, 1608 reprint of 1565 edition

Description of Ireland, Richard Stanihurst, 1577

Il Cavalerizzo, Claudio Corte, 1562, translated by Bedingfield, 1584

Britannia, William Camden, 1586

Discource of Horsemanshippe, Gervase Markham, 1583, reprinted 1584–1617

A Health to the Gentlemanly Profession of Serving Man, Gervase Markham, 1588

A Treatice of Ireland, John Dymmock, 1588

Cavalarice, Or the English Horseman, Gervase Markham, 1617 reprint of the 1607 edition

The Perfection of Horsemanship, Nicholas Morgan, 1608

Hipponomie, Michael Baret, London, 1618

Browne–His Fiftie Yeares Practice, London, 1624

Nouvelle Methode et Invention Extraordinaire de Dresser Les Chevaux, Duke of Newcastle, 1657

A New Method and Extraordinary Invention to Dress Horses, Duke of Newcastle, London, 1667

The Life of William Cavendish, Duke of Newcastle, Margaret, Duchess of Newcastle, 1667

Supplement of Horsemanship, Sir William Hope, Edinburgh, 1686

Racing Calendars, John Cheny, Ponds, James Weatherby, 1727–1773

The Sportsman's Dictionary, London, 1735

The Mercury, Stamford, Leicestershire, May 10, 1746

Dissertation on Horses, William Osmer, 1756

Treatise on Diseases and Lameness of Horses, William Osmer, 1761

Introduction to a General Stud Book, James Weatherby, 1781

Turf Register, William Pick, 1786, 1803

Sporting Dictionary, William Taplin, 1803

Privy Purse Expenses of King Henry VIII, edited by Nicholas Harris Nicholas, London, Pickering, 1827

Calendar of State Papers, Venice and Northern Italy, 1884

History of Newmarket, J.P. Hore, 1886

General Stud Book, Volume I, 1881

Old Newmarket Calendar, J.B. Muir, 1882

Breeding Racehorses by the Figure System, C. Bruce Lowe, 1885

Frampton and the Dragon, J.B. Muir, 1885

History of the Irish Horses, Michael F. Cox, M.D., Dublin, 1887

History of the English Turf, Sir Theodore A. Cook, London, 1901

Eclipse and O'Kelly, Sir Theodore Cook, 1807

Old Celtic Romances, P.W. Joyce, Dublin, third edition, 1807

A Social History of Ireland, P.W. Joyce, Dublin, 1820

Bibliography

Early Records of the Thoroughbred Horse, C.M. Prior, 1824

History of the Racing Calendar, C.M. Prior, 1826

The Roanoke Stud, Fairfax Harrison, Richmond, Virginia, 1830

Royal Studs of the Sixteenth and Seventeenth Centuries, C.M. Prior, 1835

Thoroughbred Racing Stock, Lady Wentworth, 1838

Origin and History of the British Thoroughbred Horse, J.B. Robertson, 1840

Northern Turf History, Fairfax–Blakeborough, 1848

A Bibliography of Gervase Markham, E.N.L. Poynter, 1862

The History of Horse Racing, Roger Longrigg, 1872

The Great Breeders and Their Methods, Abram S. Hewitt, 1882

The Colonial Quarter Race Horse, Alexander Mackay–Smith, 1883

The Horse Trader of Tudor and Stuart England, Peter Edwards, 1881

A Tour Through the Whole Island of Great Britain, Daniel Defoe, 1724–1726

Glossary

Akhal Teke
Horse: The best known modern descendants of the seventeenth–eighteenth century Turcoman horses (q.v.).

Aleppo
Seventeenth century: Principal "oriental" trading center, frequented by Turks as well as Arabs, near the northeast shore of the Mediterranean Sea. There was a thriving horse market. Overseas export of pure Arabian horses from this market center was forbidden by Ottomans from about 1518.

Amble
Any four-beat gait between the trot and pace. The running walk, fox-trot, rack and paso are modern examples. Required for comfort in a riding horse before the mid-eighteenth century discovery of "posting" to the trot.

Arab
Person, seventeenth century: An Arabic-speaking person, presumed member of a nomadic or pastoral tribe of the Orient (q.v.).

Arabian
Horse, seventeenth century: Horse bred by any Arab tribe, settled or nomadic, or a horse descended from or having a fraction of such ancestry. Those with the most racing (middle distance) prowess were the Persian Arabians, from the grasslands north and east of the Fertile Crescent.

Bald
Horse color marking. Facial white mark covering most of face including one or both eyes. Often with one or two blue eyes. Found on modern Welsh Mountain Ponies, Clydesdale horses and others; described on seventeenth century Running-Horses (e.g., Old Bald Peg).

Barb
Horse, seventeenth century: Any horse from North Africa, pedigree unknown or uncertain to English breeders.
Modern: Horses of Morocco and Algeria having distinct physique and genetic identity.

Bedouin
Person/tribe: Nomadic tribes of Arabs occupying the harsher deserts south of Mesopotamia. Had few horses until nineteenth century.

Bloodline
Seventeenth century: Genetic trace of racing ability back to or down from an eminent progenitor.
Modern: Trace to an ancestor.

Bred (well-bred, best-bred)
Seventeenth century: Descriptor of a stallion based on the quality of his get (e.g., He was a well-bred horse = he sired good horses.)
Modern: Quality of ancestry (e.g., He was a well-bred horse = his near ancestors were renowned.)

Breed
Seventeenth century: Horses (livestock) produced by a breeder and on the property

Classic Distance Speed
Racing capacity at 1 to 2 miles, combining sprinting and middle distance speed (q.v.), optimizing muscle chemistry, mechanical efficiency and cardiovascular capacity.

Course
Racing, seventeenth century: Speed contest among horses decided in a single running. Compare to "heat" (q.v.).

Dun
Horse color: Coat with definite sandy to clay-colored hue conferred by genes diluting basic chestnut/bay/brown coats. Common in many breeds/strains of horse, but not found in pure Arabians or in modern Thoroughbreds.

Foreign
Horse, seventeenth century: Import of any strain, lineage or use, but especially breeding stock.

Galloway
Horse, seventeenth century: 1. A small, wiry horse of the English Midlands/Yorkshire, possibly the basis, with the Irish Hobby (q.v.) of the Running-Horse (q.v.). 2. Any light horse under 14.2 hands (pony).

Heat
Racing, seventeenth century: One of a series of speed contests among horses to establish a winner of two. If after three heats no horse had won two, the three heat winners competed in a "course" (q.v.) to decide.

Hobby
Horse, 1000 BC to 1665 AD: Small, refined, ambling (q.v.) sprinter bred and raced in Ireland. Origin uncertain. Probably a local, one-time ("point") mutation prized and preserved for its sprinting (quarter-mile) prowess. Coveted diplomatic gifts from ninth through seventeenth centuries. Never established/flourished elsewhere. Key progenitor of English Running-Horse (q.v.) and American Quarterhorse.

Horse Racing
"Sport of Kings." Privilege and perquisite of rich and powerful in every horse-oriented society since before 2000 BC. Reached its zenith in England in the eighteenth century and thereafter spread throughout the world through their Thoroughbred.

Hunter
Horse, seventeenth century: A heavy, tall horse used by nobility in pursuing the hart and stag. Not a contributor to Thoroughbred blood or speed.

Jennet (Jinnit)
Seventeenth century: A small, ambling (q.v.) riding horse from Spain, or of the type seen/bred there.

Lineage
Genetic roots or ancestral basis of a horse or group of related horses.

Match
Seventeenth century: Speed contest between two horses on parallel paths, often railed to keep the contestants straight and separate.

Middle Distance Speed
Elite racing capacity over several miles, as contrasted to sprinting over ¼ to ½ mile. Prowess conferred by efficient reaching/extended gallop and high cardiovascular capacity/efficiency.

Orient
Seventeenth century: Countries/lands east of the Mediterranean Sea—little known or understood, mysterious, exotic, awesome.

Pad (Jennet)
Seventeenth century: Ambling riding horse, the type preferred for comfort over a distance.

Race
Seventeenth century: Broodstock, especially horses genetically similar, belonging to or controlled by one breeder (e.g., of his race).
Modern: 1. Contest of speed. 2. Human grouping by color, heredity or ethnicity.

Racehorse
A horse bred and kept for racing. Elite of horse breeders' art and interest for 4,000 years. Basic strains are: 1. Sprinters, originating in Ireland before Christ; 2. Middle distance (2 to 10 miles) developed in middle-east; 3. Post-seventeenth century mixture of 1 and 2 resulting in Thoroughbred and its derivatives.

Running-Horse
Seventeenth/eighteenth centuries: Sprinting racehorse of England. Mixture of Irish Hobby (q.v.) and native light horses, mostly small. (See "Galloway.")

Sprinter
Horse: Racehorse excelling at ¼ to ½ mile. All descended and deriving speed from (Irish) Hobby (q.v.).

Stone
English measure of weight. Equal to 14 lb avoirdupois.

Strain
A grouping of horses of similar type, use or talent having similar genetic makeup, especially when bred by the same person or regional group of breeders.

Thoroughbred
Term coined about 1750 to denote a racehorse (English) whose pedigree was completely known and of proven racing stock. Applied in this book retroactively to include predecessor generations of sprinter and "oriental" crossbreds between Spanker (c. 1670), the "first Thoroughbred," and Eclipse 1764. The modern Thoroughbred is the world's fastest horse at ½ mile up to 10 miles or more. The modern racing Quarter Horse is up to $^{63}/_{64}$ Thoroughbred. "Warmbloods" and nearly all other "sport" horses carry a proportion of Thoroughbred blood.

Tryall
Seventeenth/eighteenth centuries: A contest of speed between horses. "Race" is the modern term.

Turcoman
Horse, seventeenth/eighteenth centuries: Light, speedy horse bred by one of the Turkic tribes in the environs of the Black and Caspian Seas.

Turcoman Arabian
Horse, seventeenth/eighteenth centuries: Diplomatic gift horses from the stables of and/or bred by the Ottoman sultans presumed to contain both Turcoman and Arabian racing bloodlines.

Turk
Horse, seventeenth/eighteenth centuries: Any horse obtained/imported from or through Turkey, regardless of pedigree, known or unknown, or bred on English soil and descended from such an imported horse.

Turk
Person, seventeenth century: Any person from or representing the Ottoman Empire, not specifically identified with a tribal name or ethnicity. More broadly, anyone speaking a Turkic language, including Turkmen.

Yellow
Horse color, as in D'Arcy's Yellow Turk. Coat resembles polished gold. Color is peculiar to Turcoman horses from the environs of the Black and Caspian Seas, of which the best-known modern descendant is the Akhal Teke (q.v.).

Index

Adrianople 124
Ahmed III, Sultan 130, 148
Akhal-Teke 127
Aldburgh 56
Aldby Park 48, 99, 100
Aleppo xxi, 9, 48, 96–98, 100–102, 122, 124
Alexander, Robert Aitcheson 147
Almanzor 11, 100
Alvarez, Don Francisco 119
Amelia 141
American Quarter Horse 23, 31, 183
Americus 16, 168
Americus Girl 168, 171
Ancaster, Earl of 121
Andrews, Joseph 136
Angelica 61
Anne, Queen 71–72, 87, 89–90, 130
Anthony, St. 78
Antioch 98
Antoine, Chevalier de St. xx, 77–78
Arabian xiv, xvii, xx–xxi, xxv–xxvii, 2, 5–6, 9–15, 18, 35, 40, 42, 44–50, 69, 89, 92–105, 112, 114–115, 117–119, 122, 124, 127, 129–145, 147–150, 152–154, 158–159, 163–165, 170, 173
Ard Patrick 51
Ariosti 26
Arlington, Lord 120
Armagh, Bishop of 22, 28
Articles of Racing 109, 115, 164
Ashton, Sir Ralph 47
Asil 119, 122

Asti 23
Austria, Archduke of 29
Ayr 56

Babraham 133, 140–141
Baghdad 98
Bajazet 133, 140–141
Bakewell, Robert 13, 149
Bald Galloway 118
Baldwin, "Lucky" 16, 168
Baldwyn, R. & R. 140
Ballinasloe 20
Bandy 141
Baret, Michael 67, 69
Barlow, Francis xx, 67–68
Bartlett xxv–xxvi, 10, 50, 99, 101, 142, 165
Bartlett's (Bleeding) Childers xxv, 10, 50, 99, 101, 142, 165
Baxter, Nicholas 124
Bay Barb 47, 95, 112, 118–121
Bayne, Roger 34
Beach, Thomas xxi, 157
Beale, Mary xx, 88
Bedale 17, 87–88, 108, 125, 137
Bedingfield 57
Belgrade Turk 129, 147
Bellycys, Richard 55
Belmont Stakes 116, 142
Belvoir Castle 2, 34, 40, 51, 62
Berkeley, Governor Sir William xix, 30–32
Berwick, Duke of 130
Birago, Biasio de 24

Black Barb Mare 17, 51, 91
Blacklegs 102
Blackstone 67, 69
Bleeding Childers
    see Bartlett's (Bleeding) Childers
Bloody Shouldered Arabian xxi, xxv, 9, 103–104
Blunderbuss 91
Blundeville, Thomas xx, 20–21, 57, 59, 115
Bogley, Samuel E. 46
Bold Ruler 16–17, 92, 158, 168, 170–171
Bolsover Castle 75
Bolton Sweepstakes 104
Book of Falconry 72
Book of Leinster 20
Booth, Captain Richard 124
Boston 147
Boultbee, John xxi, 156
Bourdin, Ecuyer Pierre 77
Brackley 56
Bradyshe, Sir Robert 124
Brehon Laws 19–20
Brill, The 124
Brocklesby Betty 47, 50, 121
Brocklesby Park 12, 47, 121, 127
Broeck, Richard Ten 116
Brogue, Baron de 147
Brown Betty 11, 51
Buckingham, Duchess of xxiv, 37, 84, 119
Buckingham, Duke of xix–xx, xxii, xxiv–xxv, 2–3, 5, 12, 17, 34–38, 42–43, 78, 108, 125–127, 160
Buckingham, First Duke of xix, xxii, xxiv, 2, 34–36, 38
Buda, Siege of 128, 130
Bulle Rocke 99
Bullocke, James 88
Burghley, Lord 34

Burgundy, Duke of 24
Burton, Mr. 92
Burton Barb 92
Burwell Park 130
Bustler Mare 92
Butler 77
Byerley, Captain Robert xxi, 126, 128
Byerley Turk xxi, 5, 13, 45, 112, 117, 126, 128–129, 149, 153, 155, 158, 164
Byk Be Zar 12, 124
Byram, Count 118, 120
Byram Hall 45

Cade 133, 136–137, 140–141, 154
Camden, William 56, 108
Cameron, Third Baron of xix, 41
Canterbury, Archbishop of 40, 82, 119
Carew, Sir Nicholas 28
Carleton, Thomas 8, 67, 69, 81, 82, 92
    see also Carlton, Maister
Carlisle 56
Carlton, Maister 8, 67, 69, 81, 92
    see also Carleton, Thomas
Carr House 48–49, 100
Carr, Sir Robert 110
Castle Matrix xix, 30
Castle Mattress xix, 30–31
Castleton Stud 16, 168
Cato 141
Cavalarice, Or the English Horseman 23, 58, 61, 63, 64–65, 67
Cavendish, William 74, 78, 86, 107
Cavriani, Carlo 26
Chandos-Pole 88
Charles, Prince 77–78, 107

Charles I xx, xxii, 3–4, 7–8, 10, 17, 40, 72, 74, 77–78, 80–81, 84, 92, 108, 118–119, 154, 170, 172
Charles II xxiii, 2–5, 7, 10, 12, 42, 44, 50, 56, 68, 80, 84–86, 88–89, 107–111, 113, 118, 120, 122, 125–127, 147, 153, 163, 172
Charles V 28
Charles VI 148
Charlwood, John 57
Chedworth, Lord 137
Cheny, John 5, 40, 86, 115, 126, 133, 136, 147
Chester 56
Child 3, 35, 43, 50, 84, 91, 120
Childers, Leonard 9, 11, 48–49, 100
Chomondeley, Lord 139
Cistercians 55
Civil War xxii, 7–8, 74, 83, 85, 108, 116, 172
Cleveland, Vale of 108
Cleveland Bays xiv, 108
Cleveland Hills 44
Clumsy 47, 50
Cockermouth 74
Coke 12, 132–137, 147–150, 164
Conari, King 2, 19
Coneyskins 130
Conquerour 77
Constable, A. 22
Constable Burton xx, xxiv, 85, 92, 147
Constantinople 96, 98, 130, 148
Cook, Sir Theodore 28, 33, 159
Corpus Christi College 34
Corte, Claudio 57
Councillor, Aulic 24
Covenant of Kloster Seven 151
Cox, Michael F. 19
Cranbourne Lodge 152, 158
Craon, Prince de 147

Cream Cheeks XXVI, 48
Croft, Christopher 90
Croker, Richard 16, 168
Cromwell, Oliver XXII, XXIV, 2, 4–5, 7–8, 31, 41, 82, 84, 108, 119, 124–125, 132, 148, 164, 173
Cromwell, Thomas 53–55
Croyden 56
Culloden 151–152
Cumberland, Duke of XVII, 13–15, 92–93, 129, 132, 138–143, 151–161, 165
Cumberland Lodge 14, 138, 152
Cumberland Stakes 47, 105
Curwen, Henry 47, 118, 121
Curwen's Bay Barb 47, 95, 112, 118–121
Cypron 13, 154, 156

D'Arcy XX, XXII, XXIV, XXVII, 3, 5–6, 8, 12, 14, 17, 42–44, 47, 54–55, 84–95, 102, 108, 118, 120, 122, 125–127, 130–131, 137, 145, 147, 149, 160, 166
Dallugo, Count Ludovico 24
Damascus 98
Darley Arabian XXI, XXVI, 9–11, 14, 18, 48–50, 93–95, 99–102, 104, 118, 122, 142, 145, 149, 153, 159, 165, 173
Darley, John Brewster 99
Darley, Richard 100
Darley, Thomas 9, 99–100
Dedannians 19
Defoe, Daniel 104
Delavel, Sir Ralph 74
Denmark, Princess Anne of 71–72
Derby, Lord 168
Derby, Sixteenth Earl of 16
Derbyshire 12, 45, 48, 136, 148

Desmond, Fifteenth Earl of 29
Desmond, Thirteenth Earl of 23
Dettingen, Battle of 151
Devonshire (Flying) Childers XIX–XX, XXV–XXVI, 9–11, 48–51, 99–101, 113, 122, 136, 142, 147, 149, 165
Devonshire, Second Duke of 10, 45, 48, 51, 91
Diepenbeke XX, 75–76
Digby, Sir George 78
Dinsdale 5, 125–126, 128, 164
Diomed XXI, 113–114
Discource of Horsemanshippe, A 3, 57–58, 61–62, 66, 110
Dishley 13, 149
Dodsworth 118–120
Dodsworth, Dam of 120
Dodsworth, Son of 120
Domino 16, 168
Doncaster 11, 48–49, 51, 56, 100, 113, 165
Dormello 169–170
Dormouse 140–141
Dorsett Ferry 68
Drummond, A. 97
Dumplin 154
Dun Arabian XXV, 9, 101–102, 149
Durham XXI, 5, 56, 90, 104, 125, 128, 137
Durham, County XXI, 5, 90, 125, 128, 137
Durham, South County XXI, 125, 128
Dyck, Anthony Van 77
Dymmok, John 22
Edinburgh 22, 67
Edward IV V, 24
Edwards, Peter 28
Eglintown, Lord 141

Egypt 10, 69, 119
Elbe, River 151
Elizabeth I, Queen 23, 30, 57
Elizabeth II, Queen 79
Elliott 110
Epsom 14, 51, 83, 113–114, 142, 159, 165
Epsom Derby 51, 83, 113–114, 142, 165
Epsom Oaks 113
Essex, Earl of 56, 61
Euphrates 97
Eure, River 53
Eworth, Hans XXI, 122

Faber, John XXI, 132–133, 139
Fairfax, Third Lord XIX, XXII, XXIV–XXV, 2–3, 8, 37, 39–43, 82, 84, 140, 160, 172
Fairfax-Blakeborough, Major John 33, 56
Faithorne, William XIX, 41
Falmouth, Viscount 16, 166
Federation Equestre Internationale 98
Fenwick, Sir John XX, XXII, XXV, 3–4, 8, 17, 72–74, 80–82, 84, 86, 92
Fenwick, Sir William 73
Fenwick, Sorrell 82
Fenwick Barb 118
Ferrara 23
Ferrara, Duke of XIX, 24–26, 28
Ferrybridge 45
Field, The IX, XIV, 21, 90
Fielding, Lord 34
Figure System 17, 83–84, 170
FitzGerald, Gerald XIX, 28–29
FitzPatrick, Barnaby 28
FitzStephen, William 33

Flying Childers
　　see Devonshire (Flying) Childers
Flying Fox 102
Fort Worth 38
Foure chiefest Offices belonging to Horsemanship, The 21, 57, 59, 115
Fox 47, 50
Frampton 35, 37
Fulartach 20

Galteresse, Forest of 2, 33, 56–57, 109
Galtres, Forest of
　　see Galteresse, Forest of
Gatherley Moor 57
Genett 22
　　see Jennet
George I 90
George II XXIII, 13, 113, 151, 160
George III 113
Gervaix 54–55
Gervayes 54
Gladiateur 51
Gloucestershire, Earl of 20–21
Godolphin, Earl of 131, 135–137, 140, 142
Godolphin, Lord 118, 136–138, 143, 150
Godolphin, Second Earl of 131, 135, 137
Godolphin Arabian XIV, XVII, XXI, XXV, 2, 6, 12–15, 94–95, 114–115, 129, 131–145, 147–150, 152–154, 158, 164–165, 170, 173
Gog Magog 131, 134, 136–138, 141
Gonzaga, Francesco 26
Gonzaga, Marquess Federigo 26
Gonzaghesche, Le Razze 26
Grand Vizier 78

Gravesend 124
Gray Dellaval 69
Great Progenitors 13, 40, 121, 129–130, 153
Green Spring XIX, 30–32
Greenwich 27–29, 61, 71, 77
Greenwich Palace 27, 29, 71, 77
Grey Barb 92, 118, 122
Grey Markham 33
Grey Robinson 137
Grey Royal 91, 118, 137
Grey Whynot 84, 118
Grimalkin 139
Grisone, Federico 57
Guildford 113
Gwynn, Nell 110

Haddington 56
Haigh, A. G. XXI, 169
Halle 28
Hambletonian 145
Hamilton, Duke of 78
Hampton Court 5, 120, 122, 125, 130
Hannover 151
Harleian Manuscript 20–21
Harleston 56
Harley, Nathaniel 9, 101–104
Harley, Lord Edward 101–103
Harrison, Fairfax 140
Harrison, Nathaniel 112
Hartley 91, 130
Haubin 20
Hautboy 86, 130
Havresac II 16, 169–170
Heber, Reginald 133, 138, 140
Helmsley XXIV–XXV, 2–3, 5, 8–9, 12, 17, 33–35, 37–44, 46–48, 50–51, 55, 62, 82, 84, 91–92, 100–101, 108, 118–120, 125–126, 144, 149, 154–155, 160, 170, 172

Helmsley Turk 5, 42, 120, 126, 149
Hennigan 28
Henri IV 77
Henry, Prince 77
Henry VI 24
Henry VII 24
Henry VIII XIV, XIX, XXII, 2, 22, 26–29, 53, 55
Henyngham, George 28
Heremon, King 19
Hermit 51
Hewitt, Abram S. VII, 16, 166–167, 170
Highflyer XXI, XXV, 14, 153, 155–158, 164
Highflyer Hall 156, 158
Hobby III, XVII, XIV, XIX, 1–9, 14–15, 17, 19–31, 33–35, 37, 39–43, 45, 47, 49–51, 62, 63, 66, 67, 71–72, 74, 76, 78, 81–82, 84, 91–92, 101, 115, 124–126, 142, 144, 147, 149, 153–155, 158, 163, 165, 172
Hobelarii 20
Hobgoblin 134–136
Hoboy 86
Hoby, Scottish 20
Hogmagog 134, 141
Holbein XIX, 29
Holderness, Earl of 130
Holderness Turk 129–130
Hollar, Wenceslas XX, 67
Holles, Henrietta Cavendish 101
Holles, John 86, 91
Honthorst XIX, 36–37, 39
Hope, Sir William 67
Hore 28, 40, 78, 110
Hornby Castle 85, 135
Houghton Hall 138–141
Huntington 56
Hutton XXI

Index

Hutton Grey Barb 92, 122
Hutton, John 14, 91–93, 122, 158
Hutton, III, John 14, 92–93, 122, 158
Hutton, Sr., John 92, 122
Hyde Park 56
Hyperion 146
Ipswich 100
Ireland, Viceroy of xix, 28–29
Ireton 90
Isabella 26
Ishmael, Muly 120
Isle of Wight 81

James I xxii, 2–4, 9, 34, 36, 71, 73–74, 77–78, 81, 96, 172
James II, 44, 89, 121
James River 112
Jamestown xx, 30–32
Jennet 69
   see Genett
Jervaulx 53–55
Jervaulx, Abbey of 53
Jessop, William 91
Jockey Club 152, 168
Jovius, Paulus xiv, 22
Joyce, P. W. 19
Julian, Gregory 81

Keene, Foxhall 16, 168
Keene, James R. 16, 168
Kennedy, Edward 168
Kentucky Derby 51, 116
Kerouaille, Louise de 110
Kildare xix, 2, 19–20, 28–29
Kildare, Curragh of 2, 19
Kildare, Lord 28
Kildare, Ninth Earl of xix, 28–29
Kildare, Tenth Earl of 28

King Herod xxi, xxv, 13–14, 40, 121, 129–130, 132, 153, 155–159, 164, 168
Kingston 47
Kingston, Duke of 47
Kiplingcotes 56
Kirby Wharfe 44–45
Kite, John 28
Kloster Seven 151–152

Lanark 56
Lath 134, 136–137, 140–141
Latrobe xix, 32
Layton Barb 92
Layton Barb Mare 92
Le Superbe 76
Leedes xx, xxiii–xxvii, 8–9, 11, 44–50, 99–101, 127–128, 142, 155, 165
Leedes, Betty xxv–xxvi, 9, 11, 48–50, 99–101, 142, 165
Leedes, Edward xxiii, 8, 44–48, 99–100, 127–128
Leedes, Robert xxiii, 47, 155
Leedes Arabian xx, xxvi–xxvii, 11, 44–47, 49, 100–101, 127
Leedes Piping Peg 47
Leeds, Duke of 135
Lefroy 54
Leicester, Earl of 57, 136
Leinster, King of xix, 20–21
Leitrim, Lord 71
Lely, Sir Peter 43
Leonardo 26
Leopold 147
Lexington x, xxi, 16, 113–114, 116, 146–147, 166, 168, 170
Limerick, County xix, 30–31
Lincoln, City of xx, 4, 34, 71, 73, 78, 113

Lincoln, County of 73
Lincolnshire 47, 121, 127, 130, 136–137
Lister, Matthew 130
Lister Turk 129–130
Litchfield 113
Little Budworth 56
Liverpool Grand National Steeplechase 144
London Gazette 109
Longford Hall 12, 136, 148
Longrigg, Roger 138
Lonsdale, Earl of 112
Lonsdale, Lord 47, 100, 112
Lorrain, Prince of 147
Lorraine, Duke of 6, 12, 147–148
Louis XIII 71–72
Louis XIV 10, 118
Louisiana 116
Louth 130
Lowe, Bruce 17, 83–84, 91–92, 101–102, 117–118, 120–121, 137, 154, 170–171
Ludovico 24–25
Luneville Academy 147–148

Machomilia 76
Mackay-Smith 11–v, ix–x, xiii–xiv, 88
Macmurchada, Art xix, 20–21
Madrid 25
Major, John 22
Manners, Catherine xxiv, 34–35, 37–39
Manners, Francis xix, xxii, 34–35
Mantua, Marquess of 26
March, Lord 141
Markham, Gervase xx, 3, 23, 57–58, 60–62, 67, 81, 110
Markham, John 96

Markham, Robert 61
Markham Arabian 9, 95–96
Marseilles 69, 117
Marshall, Richard 121
Marske x, xx–xxi, 14–15, 91–94, 122, 144, 150, 158, 160
Marske Hall xxi, 94
Martin 141
Massey 17, 51, 91
Matchem 40, 118, 120–121, 129–130, 136, 141, 153
Mellon, Paul ix, 72, 144–145
Mercury, The 13, 149
Mereworth Castle 16, 166
Middleburg x, 59–60, 68, 75, 80
Midridge Grange xxi, 128
Milan 24
Milan, Duke of 24
Milbanke, R. 44
Miss Slamenkin 102
Mixbury 120–121
Modena 23
Modena, Archives of 24
Mohammed IV, Sultan 5, 122, 124, 132, 148, 164
Monkey xxi, 112
Monmouth, Duke of 110
Monteverdi 26
Morea 124
Morgan, Nicholas 22, 69
Morier, David 13, 132–133, 138–139
Morocco, Emperor of 3, 81–82, 84, 89, 92, 119–120, 172
Morocco, King of 92, 120
Mouvette 119
Moytura, Second Battle of 19
Muir 34–35, 37, 118
Mumtaz Mahal 14, 129, 158, 164, 168, 171

Nasrullah 146, 158, 168, 171
Native Dancer 51
Natural Barb 92, 117, 120
Natural Barb Mare 92, 117
Navesmire xxi, 112
Neapolitan Courser 61, 76
Nearco 16, 168, 170–171
Nettle Court 50
New Method and Extraordinary Invention to Dress Horses see Nouvelle Methode et Invention Extraordinaire de Dresser Les Chevaux
Newcastle, Duchess of 107
Newcastle, Earl of xxiv, 7–8, 72, 74, 78
Newcastle, First Duke of 101, 167
Newcastle Turk 91
Newmarket xxi, 3, 5, 7, 11, 28, 34, 42–45, 48, 50–51, 56, 68, 71, 74, 78, 83–84, 90, 100, 105, 109–111, 113, 118, 126, 133, 136, 163
Newmarket Heath 56, 68, 71, 110, 113
Nicolas, Nicholas Harris 27
Norfolk, Duke of 93
North Milford xx, xxiv, xxvii, 8–9, 44–45, 47, 100, 127, 155
North Milford Stud ii–iii, xxiv, xxvii, 8, 44, 47, 155
North Sea 151
Northamptonshire 118
Northumberland 3, 69, 72, 74–75
Northumberland, Earl of 3, 69, 72, 74
Northumberland, High Sheriff of 75
Nottingham 113
Nottinghamshire 57

Nouvelle Methode et Invention Extraordinaire de Dresser Les Chevaux 4, 10, 64–65, 73–77, 96, 107, 122
Nun Appleton 3, 40, 43, 84

Oakley, John 160
Obino 20
Ogle, Thomas 27
Oglethorpe, Sutton 12, 120
Oglethorpe Arabian 12
Old Careless xxv–xxvi, 9, 48, 100
Old Child Mare 120
Old Dominion 31
Old Ebony 51
Old Merlin 154
Old Rowley 110
Old Whynot 118
Onslow, Lord 141
Ormand, Ninth Earl of 28
Osmer, William 6, 132, 137, 139, 143–144, 146, 149
Ossorians 20
Ossory 20, 28
Ottoman Empire xxi, 96, 98, 122, 149
Oxford 9, 29, 57–58, 101–104, 117, 149
Oxford, Earl of 101
Oxford, Lord 9, 101, 104
Oxford Dun Arabian 101–102, 149

Paisley 56
Palio Racing 23, 26
Panton 137
Paragon 76
Parsons, Sir J. 118, 121

Patrick, St. 19
Peacock 77
Pelham, Charles 12, 42, 47, 121, 127
Pelham Bay Arabian XXVII, 12, 42, 127
Pera 10, 122
Persia 98
Persian Gulf 98
Persimmon 102
Persse, Atty 16
Pet Mare 91–92
Petworth 104
Phalaris 16, 168, 170–171
Pharos 168, 171
Phillip III of Spain 72
Phoenix 140
Pick, William 14, 47, 121, 133–134, 156
Pickering 27
Pierson, Dicky 120
Piping Peg 47, 50
Place, Rowland IX, XIV, XXIII, XXV, XXVII, 2, 5–6, 12–13, 16, 23, 42, 45, 53, 78, 89, 95, 98, 102, 104, 115, 124–129, 132, 147–149, 154–156, 158, 160, 164, 166, 173
Pluvinel, Antoine de 77
Portland, Duke of 16, 76, 101, 167
Portland, Sixth Duke of 16
Portsmouth, Duchess of 110
Poynter, E. N. L. 57
Preakness Stakes 116
Puppey 69
Puppie 67, 69

Quainton 48
Quarter Horse 23, 31, 88

Racing Calendar 5, 40, 83, 86, 105, 113, 115, 126, 133, 139, 147
Radford, R. H. 135
Ramsden, Sir William 45, 117, 128
Ranger 152
Rassegna Contemporanea 26
Rathkeale XIX, 30
Ratto, Giovanni 26
Ribot 16, 170
Richard I, King 33
Richard II, King 21
Richards, Keene 168
Richmond 33, 44, 56–57, 62, 85, 122, 125, 140
Rider, Captain 118
Roanoke River Valley 112
Roanoke Stud 112, 140
Robber 104
Roberts, Henry 140
Roberts, James 140
Robertson 51, 117, 138
Roi Herode 168, 171
Romance of Maildune 19
Romano 26
Rome 23
Rosseto 24
Roundhead 136
Roundheads, Protestant 5, 7
Rourke 71
Routh, Cuthbert 92
Roxana 134, 136
Royal Fleet 89
Royal Mare 42, 84, 86, 91, 102, 118, 120
Royal Sorrill Mare 102
Royal Stables 71, 87, 125
Royal Stud III, 3–4, 8, 17, 35, 40, 72, 74, 78, 82, 86, 88, 92, 120–121, 124, 161, 170
Royston 34

Rubecan 76
Rubens, Peter Paul XIX, 38
Running-Horse XX, 1, 3–5, 7–8, 14–15, 17, 42, 56–57, 63–69, 73–75, 77–78, 81–82, 84, 91–92, 110, 124, 144, 147, 149, 153–154, 158, 163, 165–167, 172
Rutland 17, 82, 170
Rutland, Duke of XIX, 11, 34, 51, 130
Rutland, Earl of XIX, XXII, 2, 33–35, 37–38, 40, 50, 55, 59, 62, 84, 154
Rutland, Fifth Earl of 34, 59
Rutland, Sixth Earl of XIX, XXII, 34, 35, 38, 84

Salisbury 56, 82
Salisbury, Lord 34
Salvatico, Francisco 24
Sartorius, Francis 140, 149
Saxe, Marshall 151
Scampston 154
Scanderone 102
Scanderoon 9
Scotland, King of XX, 3, 20, 71–72, 164
Scott, John 132
Secretariat III, 147, 158, 166, 168, 170
Sedburgh, Adam 54
Sedbury XX, XXIV, XXVII, 8, 14, 17, 54, 85–89, 91–95, 108, 125–126, 131, 136, 144 145, 149, 154, 160, 166
Sedbury Stud III, 8, 17, 54, 88–89, 91, 108, 125, 131, 144, 160, 166
Selima 154
Senchus Mor 19

Seven Years War 151
Seymour, James xx–xxi, 49, 90, 112
Sforza, Francesco 24
Sforza, Galeazzo Maria 24
Shaftesbury Turk 47
Shakespeare 34, 59
Shaw, William xxi, 144–145, 150
Shelley, Sir John 34
Shepherdess 141
Shotten-Herring 77
Sidely, Sir C. 141
Sidney, Sir Henry 61
Sienna 23
Simon, St. xxi, 16, 137, 146, 167, 170
Sir Archy 113–114
Skevington, Maister 28
Skewball 141
Skipjack 100
Slugg 141
Smith, Richard 57
Smithfield 14, 33, 159
Smyrna 48, 96, 98–100
Snake 130
Somerset, Duke of 100
Somerset, Sixth Duke of 104
Son of Blank xxi, 144–145, 150
Sorrill Whitenose 86
Southampton 34, 59
Southampton, Third Earl of 34, 59
Southwell, Sir Thomas 30–31
Spanish Riding School 37, 57
Spanish Succession 151
Spanker xxv, xxvii, 3, 9, 12, 42 44, 47 48, 50, 84, 100, 126–127, 130, 136, 147
Spanker Mare xxvii, 9, 42, 44, 47, 126–127, 136
Spice Islands 98
Spiletta 14–15, 158, 160, 165

Sporting Calendar 51
Sporting Dictionary x, 104, 134, 142, 152
St. James Coffee-House 133
St. Leger, Doncaster 51, 83, 113, 165
St. Quinton, Sir William 154
St. Victor Barb 118
Staffordshire 40, 72, 80, 101
Stamford 13, 56, 149
Stanihurst, Richard 22
Steven, Charles 148
Stradling Turk 130
Strickland, Sir William 48
Stuart, James xx, 3, 71–72
Stubbs xix–xxi, 15, 92–93, 114, 132, 139–140, 144–146, 149, 161
Suleiman 122
Sussex 104
Syria xxi, 96–97

Tadcaster xx, 8, 44–45, 100, 155
Tammany Hall 16, 168
Tangiers 120
Tantivy 120
Taplin, William 104–105, 133–134, 142, 152
Tara 19
Tarporley 56
Tarquin 141
Tartar 13, 155–156, 158
Tattersall, Richard xxi, 14, 156–159
Teazer 134
Tehran 98
Tesio, Federico 16, 169–170
Tetrarch, The xxi, 14, 16, 129, 146, 158, 164, 168–171
Tetratema 168
Theresa, Princess Maria 148
Thomas, Barak 16, 168

Thomas, Maister 61, 67, 81, 92
Thomas, Silken 28
Thornbery 54
Thoulouse, Count 118, 120
Thoulouse Barb 118–121
Thynne, Thomas 110
Tillemans, Peter xxi, 90, 111
Titian xix, 25–26
Toscani, Francesco 24
Tower of London 54
Tracy, Lord 100
Trajan 141
Treatice of Ireland 22
Treaty of Passarowitz 148
Tregonwell 117
Triple Crown 51, 116
Troye, Edward xxi, 146–147, 168
Tunis, Bey of 121
Tunnisland, King of 115
Turberville, George 72
Turcoman 1, 5, 12–13, 42, 115, 122, 124, 132, 149–150, 164, 172–173
Turcoman-Arabian 2, 6, 10, 13, 122, 124, 130, 145, 149, 154–155, 158, 164, 173
Turk xiv, xxi, xxv, xxvii, 2–3, 5–6, 12–14, 16, 42–45, 47, 69, 76, 84, 89, 91, 95, 100, 102, 112, 115, 117–118, 120, 122, 124–130, 132, 147–150, 153–156, 158, 164, 166, 173
Turkey 9–10, 13, 89, 98, 100–101, 104, 122, 124, 147–150, 164
Turkey, Sultan of 13, 89, 122, 148–149
Turkey Merchants 98
Turkmenistan 127
Tutbury xx, xxiv, 3–4, 8, 17, 35, 40, 72, 74, 78, 80–82, 84, 86, 88, 92, 119, 124–125, 154, 160, 170
Tutbury, Surveyor of 4, 80, 86

Tutbury Inventory xx, 4, 8, 40, 81, 92
Tutbury Royal Stud xx, xxiv, 3–4, 8, 35, 72, 74, 78, 82, 86, 88, 92, 124, 170
Two True Blues 92

U. S. Carriage Journal 149
Ubini 24, 26
Ulster, Duke of 74
Upperville 72, 144
Utrecht 36

Valentine 67, 69
Valentine, Gray 69
Valor Ecclesiasticus 53
Van Dyck 74, 77–78
Venice 25
Venus 34
Verona 23
Viceroy of Ireland xix, 28–29
Villiers, George xix, xxii, xxiv, 34, 36, 38, 78
Vintner Mare 121
Virginia, Royal Governor of xix, 30–31
Vixen 120
Voyage of Bran 19

Wales, Prince of 107
Walker, Robert 41
Wallington Stud xxv, 8, 17, 72–73, 84, 86
Walpole, Sir Robert 90
Walpole Barb 91
Walsingham, Sir Francis 61
Watkin, William 100
Watson, R. 82

Weatherby, James 47, 83, 95, 115, 126, 133–135, 142
Welbeck Abbey 8, 76, 101–103, 167
Welbeck Stud 8, 72, 75, 77, 81–82, 84
Wellesley Arabian 95
Wensleydale 53
Wentworth, Lady 10, 40, 74, 118–119, 124, 127, 150
Wentworth, Lord Deputy 40
Wharton, Lord 48, 100
White, Robert xx, 43
White Turk, The xiv, 2, 5–6, 14, 42, 45, 91, 115, 124–130, 154, 156, 158, 164, 166, 173
Whitefoot 136, 141, 148–149
Whitehall 110
Whitenose 141
Whittlebury Forest 118
Whitton Castle 90
Wicklow, County 21
Wildman, William xix, 14–15, 93, 159–160
William III, King 89, 92, 95, 121, 128
Williams, Roger 133, 136
Wilmington 58, 66
Winchelsea, Lord 10, 12, 124
Winchendon 48
Winchester 113
Windsor Castle xx, 68
Windsor Great Park 14, 138, 152
Windsor Park House 152
Woodburn Farm 147
Wootton, John xix–xxi, 11, 45–47, 90, 99, 104, 129, 138
Workington Hall 47, 118, 121
World War I 16, 168
Wormwood 5, 126

Wyvill, Isabel xxiv, 8, 87
Wyvill, Sir Marmaduke xx, 85, 92, 147

Yarborough, Earl of 123
Yellow Turk xxvii, 3, 5, 12, 42–44, 84, 91, 126–127, 130
York, City of 56–57, 112
Yorkshire, North 2, 5, 7–9, 17, 33–35, 44, 53, 55, 57, 82, 85, 87, 89, 91–92, 100, 104, 108, 120, 122, 125, 130–131, 135–137, 147, 154, 164
Yorkshire, North Riding of 33, 85